Practicing Sustainability

Practicing Sustainability

Guruprasad Madhavan
Barbara Oakley
David Green
David Koon
Penny Low
Editors

Foreword by **Michael Spence**
Editorials by **Klaus Schwab**, **Robert Rubin**, and **George Whitesides**
Afterword by **M.S. Swaminathan**

# Springer

Editors
Guruprasad Madhavan
Program Officer
The National Academies
Washington, DC, USA

Barbara Oakley
Associate Professor
Oakland University
Rochester, MI, USA

David Green
Social Entrepreneur
El Cerritto, CA, USA

David Koon
Former Member, New York State
  Assembly
Albany, NY, USA

Penny Low
Member of Parliament, Singapore

The views expressed in this book are those of the individual authors alone and do not necessarily reflect the policies or positions of their respective organizations.

–The Editors

ISBN 978-1-4899-8898-0   ISBN 978-1-4614-4349-0 (eBook)
DOI 10.1007/978-1-4614-4349-0
Springer New York Heidelberg Dordrecht London

© Springer Science+Business Media New York 2013
Softcover reprint of the hardcover 1st edition 2013
Chapter 44 was created within the capacity of an US governmental employment. US copyright protection does not apply.
This work is subject to copyright. All rights are reserved by the Publisher, whether the whole or part of the material is concerned, specifically the rights of translation, reprinting, reuse of illustrations, recitation, broadcasting, reproduction on microfilms or in any other physical way, and transmission or information storage and retrieval, electronic adaptation, computer software, or by similar or dissimilar methodology now known or hereafter developed. Exempted from this legal reservation are brief excerpts in connection with reviews or scholarly analysis or material supplied specifically for the purpose of being entered and executed on a computer system, for exclusive use by the purchaser of the work. Duplication of this publication or parts thereof is permitted only under the provisions of the Copyright Law of the Publisher's location, in its current version, and permission for use must always be obtained from Springer. Permissions for use may be obtained through RightsLink at the Copyright Clearance Center. Violations are liable to prosecution under the respective Copyright Law.
The use of general descriptive names, registered names, trademarks, service marks, etc. in this publication does not imply, even in the absence of a specific statement, that such names are exempt from the relevant protective laws and regulations and therefore free for general use.
While the advice and information in this book are believed to be true and accurate at the date of publication, neither the authors nor the editors nor the publisher can accept any legal responsibility for any errors or omissions that may be made. The publisher makes no warranty, express or implied, with respect to the material contained herein.

Printed on acid-free paper

Springer is part of Springer Science+Business Media (www.springer.com)

The test of a first-rate intelligence is the ability to hold two opposing ideas in the mind at the same time and still retain the ability to function. One should, for example, be able to see that things are hopeless and yet be determined to make them otherwise.

– F. Scott Fitzgerald

Foreword

When I first received the manuscript for *Practicing Sustainability*, I have to admit I opened it with some trepidation – fearing something ponderous, or worse. What I found instead was a delightful surprise. This is a book that captures the vast territory of adapting our values, habits, practices, thinking, systems, and technologies to enhance the quality of life on the planet we share. The chapters are gems of precision and insight written by doctors, architects, chefs, scientists, economists, social innovators and many others, each bringing different dimensions and perceptions to the idea of sustainability. The result is a rich menu that allows readers to connect through familiar experiences.

The Chinese discuss sustainability more and more these days. With their 1.3 billion people and their large and growing economy on the way to being a future economic giant, inventing a different or modified growth model is a key to their long-term development. But when people talk about sustainability in day-to-day conversation, they speak in terms of changing lifestyles, and by implication values. In a way, this is what *Practicing Sustainability* does too. It brings sustainability down to real life: how we live, what we eat, how we move around, how we distribute income and wealth.

To be sure, aspects of moving in the direction of sustainability will require policies and international understandings. But more important in the long run are the millions of innovations and adaptations, informed by a responsible sense of interdependence and a commitment to intergenerational equity, that make up the substance of learning to live in ways that are sustainable. *Practicing Sustainability* is so convincing in part because, rather than discussing the subject in the abstract, it *illustrates* it. And in so doing, it helps define it in an appropriately expansive way.

There are many fine features of this book. One of them is to make a very convincing case that in practice, sustainability involves innovations – thousands of them – and is a decentralized process: a market for ideas in effect, the best of which spread. Another is that for us, and our children and grandchildren, sustainability is a journey of discovery, not a one-time shift, or a well-defined endpoint. While there may be points at which difficult trade-offs are made, fundamentally this is an adventure, a serious one to be sure, but one where the human tendency to empathy and creativity shines. A third strength is that the book captures the diversity of experience around the world, recognizing that at this stage in history, developing countries have lower incomes and hence different aspirations than developed economies, ones that focus on growth, poverty reduction, and inclusiveness. Experience suggests that major failures on the inclusiveness front generally disrupt growth through a loss of support for the policies needed to sustain growth. In others words, some notion of balance is part of the economic social dimensions of sustainability, and that comes through clearly in the book. And it is coming back as an issue in the advanced countries too.

With so much variety in subjects, perspectives, and ideas, summarizing is a challenge. But in a way, that is really the great strength of this volume. Since the important work of Brundtland Commission, sustainability has been thought of as conducting ourselves in such a way as not to impair the opportunities of future generations. Sounds fundamentally right – keep some kind of expansive planetary balance sheet intact. It is consistent with the endowment model that universities and the Norwegian Sovereign Wealth fund use to allocate resources across time. But as the editors correctly note, the idea is a bit too abstract to connect well with individuals, businesses, and communities.

By contrast, the book defines sustainability by example, and broadens the subject in a way that brings it close to home. Finding and practicing sustainability is explicitly defined by the editors as a work in progress, as a bottom-up process in significant part. The sense of humility in the face of great challenges comes through clearly, as does the sense of adventure.

The message in the end is that sustainability is based on a core value of intergenerational equity and respect for the planet and all of its occupants. Practicing it is a multidimensional set of activities in all facets of life, and much of it comes down to innovation. If you read between the lines a little, sustainability comes through clearly as having elements of stability, balance, and equity, as well as dealing with resources not in infinite supply. Learning about complex things isn't always fun, but in this case it really is, and rewarding too.

Michael Spence

Preface

"Sustainability" means different things to different people. It is underpinned by differing – and frequently contradictory – preconceptions, public attitudes, political agendas, cultural beliefs, emotions, and goals. We customize sustainability to fit with our needs, lifestyles, and belief systems. Importantly, there is a tension between personal sustainability and global sustainability. This is an echo of the tensions between caring for oneself versus caring for one's community. Practicing and achieving sustainability starts by being willing to look critically at the concept. It also means enabling rich and vigorous discussion to determine a framework for best ideas and practices. That's what this book is an attempt to do.

In *Practicing Sustainability*, contributors pour their distilled life experiences into their essays. The writers come from an extraordinarily broad range of backgrounds: poet, symphony orchestra conductor, secular evangelical pastor, chef, skyscraper architect, filmmaker – all these, along with visionaries, scientific leaders, business executives, practitioners, entrepreneurs, policy makers, and contrarians in sustainable development. What emerges from these essays is a wide spectrum of views that confirm one thing: Sustainability is perceived and pursued in different ways not only due to different interpretations, but also because of different trade-offs, incentives, values, and tensions associated with it.

How can we understand and make the best out of these differing views? A peek back at the history of the sustainable development might help. In the 1980s, a commission led by former Norwegian Prime Minister Gro Harlem Brundtland helped kindle public awareness about sustainability. The commission's definition of sustainable development has become one of the most widely cited: "development that meets the needs of the present without compromising the ability of future generations to meet their own needs." The definition is inspirational, but difficult to implement.

So what *is* implementable? If there is one basic message in this book, it is that more of the same is not the right approach to sustainable development. Worse yet, aspiring to create a more sustainable society by simply throwing more money or finite natural resources at something without thinking through other realistic options may actually *impede* sustainable development. Increasingly sophisticated use of technology has enabled humans through thousands of years to overcome apparent resource limits. An important message of this book, however, is that technology forms only one route toward achieving sustainability. Pragmatism and common sense are also key.

Ultimately, the key questions remain: What is it we are trying to sustain? As a society, are we capable of practicing or achieving sustainable development? Is the concept of sustainable development realistic? What are our social – and personal – limits, constraints, and responsibilities? How do we resolve or take advantage of the opportunities that tension between growth and sustainability can afford us? There is no "one-size-fits-all" answer to these questions. With time and the much needed critical thinking, sustainable development will become a more integral part of our culture. We are not there yet, but we hope *Practicing Sustainability* will serve as a stepping stone.

Guruprasad Madhavan
Barbara Oakley
David Green
David Koon
Penny Low

Acknowledgments

The editorial team for *Practicing Sustainability* emerged at a recent World Economic Forum Annual Meeting of New Champions in Tianjin, China. For their help toward catalyzing our partnership, and for their guidance and mentorship, we would like to profusely thank Bruce Alberts and Klaus Schwab.

We were fortunate to work with a superb group of contributors and are grateful for their participation. Editing *Practicing Sustainability* was like organizing an *a capella* ensemble with breathtaking vocal range.

We appreciate the generous encouragement, wisdom, and support of Paul Beaton, Clyde Behney, Geeta Bhatt, Joshua Brandoff, Jason Cole, David Dierksheide, Kathryn Fletcher, Harvey Fineberg, Kevin Finneran, Claudia Grossmann, Sarah Hall Gueldner, Dennis Hartel, James Hartel, Erin Hogbin, Toinette Lippe, Margo Martin, Rose Marie Martinez, Joann McGrath, Stephen Merrill, Aparna Raja, Robert Rubin, Kinpritma Sangha, Michael Spence, M.S. Swaminathan, Charles Vest, George Whitesides, Adam Winkleman – and our families.

It was a special treat to work with David Packer and Sara Kate Heukerott at Springer – their zeal and insights helped us to march confidently toward our vision for this challenging project. We also thank Lesley Poliner and Vinita Arokianathan for their excellent work with the book production.

Biographical Information

Guruprasad Madhavan (Editor) is a program officer at the National Academy of Sciences, National Academy of Engineering, Institute of Medicine, and National Research Council – collectively known as the National Academies – in Washington, DC. Madhavan received a bachelor's degree in engineering from the University of Madras (India), M.S. and Ph.D. in biomedical engineering, and M.B.A. from the State University of New York (SUNY). His doctoral research work was focused toward developing noninvasive neuromuscular stimulation approaches for improving circulation. Among other awards and honors, Madhavan has received the AT&T Leadership Award, SUNY Chancellor's Promising Inventor Award, Rotary Foundation's Paul Harris Fellowship, Institution of Engineering and Technology's Mike Sargeant Career Achievement Award, *EE Times*' Student of the Year Award, American College of Clinical Engineering's Thomas O'Dea Advocacy Award, American Society of Agricultural and Biological Engineers' Robert Stewart Engineering-Humanities Award, Association for the Advancement of Medical Instrumentation's AAMI-Becton Dickinson Award for Professional Achievement, District of Columbia Council on Engineering and Architectural Societies' Young Engineer of the Year Award, and IEEE-USA Professional Achievement Award. Madhavan was also selected as one among 14 people as the "New Faces of Engineering" in the *USA Today* in 2009. He is an IEEE Ambassador, and co-editor of *Career Development in Bioengineering and Biotechnology* (Springer) and *Pathological Altruism* (Oxford University Press).

Barbara Oakley (Editor) is an associate professor of engineering at Oakland University in Rochester, Michigan, and a former vice-president of the IEEE Engineering in Medicine and Biology Society – the world's largest bioengineering society. She earned a B.A. in Slavic languages and literature and a B.S. in electrical engineering from the University of Washington in Seattle. She received an M.S. in electrical and computer engineering, and a Ph.D. in systems engineering, both from Oakland University. Oakley's research and teaching interests are in the area of bioelectronics, medical sensors and instrumentation, and the effects of electromagnetic fields on biological cells. Oakley is a recipient of the National Science Foundation's (NSF) Frontiers in Engineering New Faculty Fellow Award, John D. and Dortha J. Withrow Teaching Award, Naim and Ferial Kheir Teaching Award, and was designated as NSF New Century Scholar. She has also received the NSF Antarctic Service Medal following her work as a communications expert at the South Pole Station. Prior to her academic career, Oakley also rose from the ranks of Private to Captain in the U.S. Army, during which time she was recognized as a Distinguished Military Scholar. Oakley is an elected fellow of the American Institute for Medical and Biological Engineering and the author of *Hair of the Dog* (WSU Press), *Evil Genes* (Prometheus Books), and *Coldblooded Kindness* (Prometheus Books). She is also a co-editor of *Career Development in Bioengineering and Biotechnology* (Springer) and *Pathological Altruism* (Oxford University Press).

David Green (Editor) is a MacArthur Fellow, is an Ashoka Fellow, and is recognized by the Schwab Foundation as a leading social entrepreneur. Green has been the prime mover in three successful technology transfers, which have had a significant impact in the fields of blindness prevention and amelioration of hearing impairment. Green helped develop Aravind Eye Hospital in Madurai, India, known as the largest eye care system in the world. Seventy percent of the care is provided free of charge or below cost, yet the hospital is able to generate substantial surplus revenue. Green has replicated this cost recovery model in Nepal, Malawi, Egypt, Guatemala, El Salvador, Tibet, Tanzania, and Kenya and has assisted the Lions Aravind Institute for Community Ophthalmology to build their capacity to provide this assistance to well over 200 programs worldwide. In collaboration with Seva Foundation and Aravind Eye Hospital, Green directed the establishment of Aurolab (India), the first nonprofit manufacturing facility in a developing country to produce affordable intraocular lenses to ameliorate cataract, which is the main cause of blindness. Green also helped establish suture manufacturing at Aurolab. Green collaborates with the International Agency for the Blind and Deutsche Bank to create an "Eye Fund" that will improve financing for sustainable eye care; collaborates with Grameen Health in Bangladesh to develop eye hospitals; and works with California Health Care Foundation to develop affordable retinal imaging for eye disease detection and monitoring. Green also is vice president of Ashoka, where he works to develop more abundant and efficient financing for the social sector. Green received his M.P.H. degree from the University of Michigan. Among other numerous honors, he has received the Spirit of Helen Keller Award.

David Koon (Editor) represented the 135th district of the New York State Legislative Assembly from 1996 to 2010. During his service, Assemblyman Koon served as a member of the Alcoholism and Drug Abuse; Economic Development, Job Creation, Commerce and Industry; Local Governments; Small Business; and Library and Education Technology Committees. He also served as chair of the Legislative Commission on Rural Resources and vice chair of the Legislative Commission on Toxic Substances and Hazardous Waste. For his sustained leadership toward crime and violence reduction, Koon has received the Peacemaker Award from the Center for Dispute Settlement and New York State Crime Coalition's Crime Prevention Award and has also served on President Clinton's Task Force against Youth Drugs and Violence. He has received the 911 Professional Award from the E911 Institute and the President's Award of the National Emergency Number Association for his pioneering work on enhancing the E911 access in New York State. Koon is also an active board member for the National Center for Missing and Exploited Children and is the founding co-director of the Jennifer Koon Peacemaking Foundation. Upon graduation from Fairmont State College in West Virginia, Koon was employed with Kelly Springfield Tire Company and later with Bausch and Lomb as an industrial engineer for nearly 15 years. Koon is also a graduate of the Rochester Citizens Police Academy.

Penny Low (Editor) is a Member of Parliament of Singapore and was the youngest elected female representative in 2001. Currently, she chairs the Government Parliamentary Committees (GPC) for Ministry of Information, Communication and the Arts, and serves on the GPC for Ministry of Finance, Trade and Industry. In addition to being the chair of the Singapore-Peru Parliamentary Friendship Group, vice-chair of the North East Community Development Council, and the Pasir Ris-Punggol Town Council, she is also a Labour Union Advisor, active on several national committees, in particular those promoting enterprises and social entrepreneurship, town and community development, and internation parliamentary friendship groups. Low has extensive experience in the wealth management industry, having founded the Planners-Hub Consultancy. She serves on the Global Voice Editorial Advisory Board of Financial Planning Association (USA) and the ISO Taskforce on Personal Financial Planning. She trains senior executives and has developed courses for the designations of Certified Financial Planner, the Chartered Financial Consultant, the Chartered Life Underwriters designation, and an undergraduate module on Financial Management. Low is one of the founding members of the New Asian Leaders and the Forum of Young Global Leaders of the World Economic Forum. She completed the Global Leadership and Public Policy program at the John F. Kennedy School of Government at Harvard University. Low has been recognized by Yale University as a Yale World Fellow and a Yale Scholar. Low founded the Social Innovation Park – a nonprofit organization promoting thought leadership and social entrepreneurship – recognized as a leading social enterprise in Singapore. Among other boards, Low serves on the Jet Li One Foundation and the Commonwealth Business Council. She has been recognized by the Aspen Institute as an Aspen Ideas Festival Scholar and has been listed in *Singapore Tatler*'s Top 300 for many years.

Michael Spence (Foreword) is William R. Berkley Professor in Economics and Business at New York University's Leonard N. Stern School of Business. He is a senior fellow at the Hoover Institution, the Philip H. Knight Professor and Dean Emeritus of Management in the Graduate School of Business at Stanford University, and a distinguished visiting fellow at the Council on Foreign Relations. Earlier, he was a professor of economics and business administration, chairman of the economics department, and dean of the Faculty of Arts and Sciences at Harvard University. Spence, whose scholarship focuses on economic policy in emerging markets, the economics of information, and the impact of leadership on economic growth, was chairman of the Commission on Growth and Development, an independent global policy group focused on strategies for producing rapid and sustainable economic growth and reducing poverty. He also serves as a consultant to PIMCO, as a senior adviser at Oak Hill Investment Management, and as a member of the board of the Stanford Management Company as well as a number of public and private companies. He earned his undergraduate degree in philosophy (*summa cum laude*) at Princeton University and was selected for a Rhodes scholarship. He was awarded a B.S. and M.A. in mathematics from Oxford University, and earned his Ph.D. in economics at Harvard University. Spence is a member of the American Economic Association, and a fellow of the American Academy of Arts and Sciences and the Econometric Society. Among his several honors and awards, Spence was awarded the 1978 John Kenneth Galbraith Prize for excellence in teaching, the 1981 John Bates Clark Medal from the American Economics Association, and the 2001 Nobel Memorial Prize in Economic Sciences. He is author of *The Next Convergence: The Future of Economic Growth in a Multispeed World* (Farrar, Straus and Giroux).

Klaus Schwab (Editorial) is founder and executive chairman of the World Economic Forum. Founded in 1971, the World Economic Forum is the foremost global partnership of world leaders in business, government, academia, and civil society committed to improving the state of the world. Schwab holds a Ph.D. in economics from the University of Fribourg, a Ph.D. in engineering from the Swiss Federal Institute of Technology, and an M.P.A from the Harvard University's John F. Kennedy School of Government. Earlier, Schwab was professor of business policy at the University of Geneva. He is the author or co-author of several books, including the *Global Competitiveness Report* – an annual publication since 1979 that examines the potential for increasing productivity and economic growth of countries around the world. Schwab is also co-founder of the Schwab Foundation for Social Entrepreneurship and founder of the Forum of Young Global Leaders. He has received several honorary doctorates and is an honorary professor of the Ben-Gurion University of Israel and the China Foreign Affairs University. Schwab has also worked with many other organizations, including serving as a Trustee for the Peres Center for Peace, and has served as an advisor on sustainable development to the United Nations. Schwab has been listed by *Time* as one of the "100 Most Influential People" who are transforming the world. Among his numerous awards and honors are the Dan David Prize and a Knighthood bestowed by Her Majesty the Queen of England.

Robert Rubin (Editorial) is Co-Chairman of the Council on Foreign Relations. He began his career in finance at Goldman, Sachs & Company, where he later served as Vice-Chairman and Co-Chief Operating Officer, and as Co-Senior Partner and Co-Chairman. Before joining Goldman, he was an attorney at the firm of Cleary, Gottlieb, Steen and Hamilton in New York City. Long active in public affairs, Rubin joined the Clinton Administration in 1993 as Assistant to the President for Economic Policy and as Director of the newly created National Economic Council. From 1995 to 1999, Rubin served as the 70th Secretary of the Treasury, where he was involved in balancing the federal budget; opening trade policy to further globalization; acting to stem financial crises in Mexico, Asia, and Russia; helping to resolve the impasse over the public debt limit; and guiding sensible reforms at the Internal Revenue Service. From 1999 to 2009, Rubin served as a member of the Board of Directors at Citigroup and as a senior advisor to the company. In 2005, Rubin was one of the founders of The Hamilton Project, an economic policy project housed at the Brookings Institution. He is a member of the Harvard Corporation, a member of the Board of Trustees of Mount Sinai Medical Center, a counselor for Centerview Partners, and Chairman of the Board of the Local Initiatives Support Corporation. Rubin graduated *summa cum laude* from Harvard College with an A.B. in economics. He received an L.L.B. from Yale Law School and attended the London School of Economics. Rubin is the author of the *New York Times* bestseller *In an Uncertain World: Tough Choices from Wall Street to Washington* (with Jacob Weisberg; Random House, 2003), which was also named as one of *Business Week's* 10 best business books of the year. Rubin is recipient of honorary degrees from Harvard, Yale, Columbia, and other universities.

Biographical Information

George Whitesides (Editorial) is the Woodford L. and Ann A. Flowers University Professor of Chemistry and Chemical Biology at Harvard University. Whitesides' scientific contributions come from such diverse areas as nuclear magnetic resonance spectroscopy, materials and surface science, microfluidics, and nanotechnology. He is best known for his insights into surface chemistry, understanding how molecules arrange themselves on a surface. The discovery laid the groundwork for advances in nanoscience that led to new technologies in electronics, pharmaceutical science, and medical diagnostics. His recent research interests include energy, the origin of life, and science for developing economies. Whitesides is active in numerous public service roles. He has served on advisory committees for the U.S. National Science Foundation, NASA, Department of Defense, and the National Research Council. Whitesides is the author of over 1,100 scientific articles and is listed as an inventor on more than 100 patents. He is a member of the American Academy of Arts and Sciences, the U.S. National Academy of Sciences, and the U.S. National Academy of Engineering. He is a foreign associate of the Royal Society of Chemistry (U.K.), the Royal Netherlands Academy of Arts and Sciences, the Indian Academy of Sciences, and the French Academy of Sciences. He is also a fellow of the American Association for the Advancement of Science, the Institute of Physics, and the American Chemical Society. Among other honors, he has received the U.S. National Medal of Science, the Kyoto Prize in Materials Science and Engineering, the Welch Award in Chemistry, the Dan David Prize in Future Science, the American Chemical Society Priestley Medal, the Prince of Asturias Award in Science and Technology, the Dreyfus Prize in Chemistry, the Othmer Gold Medal of the Chemical Heritage Foundation, and the King Faisal International Prize. Whitesides earned a bachelor's degree from Harvard University and a Ph.D. from the California Institute of Technology, and has previously served as a faculty member at the Massachusetts Institute of Technology. He is co-author of the books *On the Surface of Things* and *No Small Matter* (Harvard University Press).

M.S. Swaminathan (Afterword) has been acclaimed by *Time* as one of the "20 Most Influential Asians of the 20th Century" and one of the only three from India – the other two being Mohandas Gandhi and Rabindranath Tagore. He has been described by the United Nations Environment Programme as "the Father of Economic Ecology" and by Javier Perez de Cuellar, Secretary General of the United Nations, as "a living legend who will go into the annals of history as a world scientist of rare distinction." Swaminathan has served as independent chairman of the Food and Agricultural Organization Council, president of the International Union for the Conservation of Nature and Natural Resources, president of the World Wide Fund for Nature (India), president of the Nobel Peace Prize–winning Pugwash Conferences on Science and World Affairs, president of the National Academy of Agricultural Sciences, and as chairman of the Indian National Commission on Farmers. He also served as acting deputy chairman and later member of the Planning Commission of India and director-general of the International Rice Research Institute, the Philippines. He was a trustee of Bibliotheca Alexandrina in its formative years. A plant geneticist by training, Swaminathan's contributions to the agricultural renaissance of India have led to his being widely referred to as the scientific leader of the green revolution movement. Among other honors, Swaminathan has been awarded the Ramon Magsaysay Award for Community Leadership, the inaugural World Food Prize, the Golden Heart Presidential Award of Philippines, Commandeur of the Order of the Golden Ark of the Netherlands, the Albert Einstein World Science Award, Ordre du Merite Agricole of France, the Sasakawa Environment Prize, the Honda Prize, Volvo Environment Prize, Tyler Prize for Environmental Achievement, the Indira Gandhi Prize for Peace, Disarmament and Development, the Franklin D. Roosevelt Four Freedoms Medal, the UNESCO Gandhi Gold Medal for Culture and Peace, and the Lal Bahadur Sastri National Award. He has received three of the top civilian awards given by the President of India – Padma Sri, Padma Bhushan, and Padma Vibhushan. Swaminathan is an elected member or fellow of many of the leading scientific academies of the world, including the Royal Society (U.K.) and the U.S. National Academy of Sciences. He has received nearly 60 honorary doctorates from universities around the world. He currently holds the UNESCO Chair in Ecotechnology at the M.S. Swaminathan Research Foundation in Chennai, India. Since 2007, he has also served as a Member of the Parliament of India.

Editorial

A Collaborative Framework for Practicing Sustainability

The state of the world today has never been more complex; we are, in many ways, entering unchartered territory – a period of extraordinary opportunities, yet nonetheless fraught with unprecedented imbalances and tremendous risks. Effectively managing our global affairs will require the creation of more productive, sustainable, and collaborative systems of cooperation which engage all stakeholders of global society, specifically, one that transcends the traditional barriers of politics and economics; brings different organizations and individuals together to form true public-private partnerships; and has the trusted organizational capability to pursue pragmatic solutions. The World Economic Forum provides this collaborative global framework.

The stakeholder concept – which I first described in a book in 1971 – is the basis for everything we do at the World Economic Forum, and it is deeply interwoven with the notion of global citizenship. It is the idea that citizens have both rights and shared responsibilities – that the pursuit of our own interests can only be substantially realized by incorporating the interests of all those with whom we have a mutually dependent relationship. This is true on all levels, and in any capacity in which we take decisions: family life, society, business, or politics.

With this background, I can say that *Practicing Sustainability* is a compelling, seminal volume. It presents wide-ranging views and approaches toward better defining, understanding, and practicing sustainability. This book explains that the concept and goals of sustainable development are themselves unclear. But it does so not by means of an abstract discourse or an entrenched mindset but rather by building on broad practical experiences of different stakeholders. All of this is catalyzed by the talents of an unusually diverse editorial team who have called on a wide range of perspectives. I very much hope that the lessons and wisdom contained in the pages of *Practicing Sustainability* will spread further and lay the foundation for the next generation of ideas for sustainable development.

The stakeholder concept is fundamental to all of this. Philosophically, it is embedded in the Confucian notion that "the world is no one's private property." In other words, we need to share what we have. I am also aware that the Western Enlightenment paradigm of prioritizing individual rights over society and of market benefits can be at the expense of strengthening overall community spirit. The views of German philosopher Georg Wilhelm Friedrich Hegel may offer a middle ground. Hegel developed the dialectic concept of ultimately integrating and uniting both Eastern and Western concepts – namely, combining individual creativity with common responsibility.

Practicing Sustainability fits well within this framework. It is a thoughtful and refreshingly different book, one that underscores the need for individual creativity coupled with common sense, critical inquiry, and shared responsibility. If we all recognize that in the world of today we share the values of the stakeholder concept – or in other words, the concept of a harmonious society, not only on a national but also on regional and global levels – then we will create a more peaceful, inclusive, and sustainable world.

Klaus Schwab

Editorial

Sustaining Sustainability: Thoughts on Managing Trade-Offs and Competing Considerations

As this book so aptly demonstrates, the concept of sustainability can be applied to many different areas, which could be woven together into a sustainability agenda. This would position sustainability as an overall, broadly applicable principle for many of our economic and societal activities. Let me give you three examples.

One that is highly visible right now is the imperative for *fiscal sustainability*, that is, for a sound fiscal regime. I'll discuss that in the context of the United States, but the same conceptual analysis would apply elsewhere, though the differences in circumstances would affect the specifics of each program. A failure to effectively redress our present unsustainable fiscal outlook is highly likely to lead to market and economic crisis at some unpredictable point – as evidenced in the sovereign debt crisis that recently developed in the euro zone. Moreover, fiscal policy is one area where, properly structured, a sustainable program could be beneficial for both the short term and the long term, though many trade-off choices would be required amongst the vast array of possible budgetary programs and revenue measures, and on the levels of spending and revenues. For both the short term and the long term, a sound fiscal program now could significantly improve business confidence, provide capacity for public investment – so critical to competitiveness and growth, create greater resilience to deal with economic downturns and national emergencies, prevent crowding out of private investment, and protect against the risk of severe crisis. Key to benefiting both the short term and the long term is enacting the full fiscal program now, including an upfront stimulus, but deferring implementation of deficit reduction, the program's central focus, for two or three years to allow recovery to take better hold. The stimulus would almost surely have substantially greater and longer-lasting effect if combined with the confidence likely to result from effectively addressing our unsustainable fiscal outlook.

Environmental sustainability, a second area, is necessary for clean air, clean water, natural habitat for quality of life, protection against the severe dangers of global warming, and much else, and can avoid the enormous expenses required to remediate an unsound legacy. A central point here is that environmentally damaging activities create social costs that should be included both in our measurements of GDP and in the way that individual environmental issues are evaluated. Then, the organizations that create these social costs can absorb them, or those costs could be dealt with through regulation. The measurement of social costs is highly imperfect and controversial, and always will be, but that can be improved over time. Meanwhile, the perfect shouldn't be the enemy of the good, as long as the process has safeguards for integrity in decision making.

A third area is *healthcare sustainability*. As medical technology improves and life expectancies increase in many countries, strong upward pressures will continue on the cost of healthcare per patient and in the economy as a whole. Healthcare will always, inevitably, fall short of demand and need, and there are no easy answers with respect to improving outcomes and constraining costs. Even mainstream healthcare analysts have very different views on these issues. With that in mind, healthcare needs a robust framework within which to evaluate different approaches – perhaps as taken by different states in our federal system – to develop best practices for a sustainable system. Here, the trade-offs may not be across time, but with respect to care versus cost and the effects of different policy choices. Even if some approaches are found that improve both outcomes and costs – and that would seem highly likely given the present state of our system – policy makers will still ultimately face these trade-off decisions, albeit from a sounder point of departure base.

These are just some of the issues that could be governed by the principle of sustainability, if we better understood that principle. That then raises the question of what sustainability actually means. The book revisits a broadly held definition: "development that meets the needs of the present without compromising the ability of future generations to meet their own needs." That definition is probably unobjectionable but is also so abstract that it does not provide much, if any, guidance for actual decision making. More specifically, it provides little direction for weighing trade-offs and other competing considerations, whether across time or within a horizontal timeframe. Substantively, that means the definition provides no concept or methodology for aiming at the objective of optimizing outcomes. In a political context, decisions that are more heavily weighted one way or the other can undermine the credibility of the sustainability principle. And decisions are easier to defend politically when rigorous optimization can be demonstrated. Moreover, sustainability is often seen – rightly in my view – as a moral imperative, but that again requires criteria for evaluating decisions. So, overall, for the concept of sustainability itself to be sustainable requires a rigorous approach to decision making.

The key to all the trade-off choices and competing considerations inherently involved in every area of sustainability is cost–benefit analysis. Properly used, it would produce optimization in decision making and a better position for defending outcomes politically. The fear is often expressed that cost–benefit analysis could be used to minimize needed regulation; and that may have been the motivation of some proponents of this approach. However, that is a question of developing processes that are designed to effectively protect the integrity and objectivity of decision making. One suggestion, at the federal level, would be to move this evaluation function to a nonpartisan, independent organization, analogous to the Congressional Budget Office, where integrity and professionalism have served our budgetary processes so well. Moreover, properly applied, this methodology would lead to increases as well as reductions, or reconfigurations in regulations, for example, with respect to the management of climate change.

The more difficult questions lay in the identification and measurement of costs and benefits – for example, the many choices in constructing a fiscal program, the social costs in the environmental area, and the healthcare trade-offs. Those difficulties are greatest when the benefits or the costs are intangible, as with peace of mind from having a social safety net, quality of life, and the adverse effects of unsound fiscal conditions on confidence. However, decisions in all of these areas and many others inevitably create costs and benefits. The only question is whether these decisions will be made with an explicit and rigorous analysis or an incoherent, nonrigorous, and probably in many instances, somewhat subjective manner. And as cost–benefit analysis is applied, measurements would improve over time. Meanwhile, again, the perfect should not be the enemy of the good in improving the outcomes for our economy and our society and in buttressing the politics of sustainability.

Robert Rubin

Editorial

Sustainability: Can You Get There, if You Don't Know Where "There" Is?

Sustainability. What *is* it? What it is *not* is a concept with a sharp definition. Its general meaning is clear: It represents a wish for a world in which human uses of resources do not produce irreversible, global-scale change, where consumption (for example, of energy) is balanced by replenishment (from the sun), and where waste (for example, carbon dioxide) does not produce harmful change (of climate). Ultimately, it is a hope for stability. It is, in many ways, more a mood or aspiration than a clear direction. The almost undefined, aspirational, or even sometimes ideological character of sustainability may be fine in giving a name to an intention: We spend much of life pursuing concepts – beauty, liberty, justice – that we are hard-pressed to define in precise terms.

I'm a scientist and an engineer, and the intention of sustainability seems to me to be self-evidently important. I also understand the laws of thermodynamics, and know that it is not useful to try to get something for nothing, at least in technologies based on energy and its uses. Technology will be only a part of any answer that leads toward sustainability, but since people – and the technologies they use to make their lives better – are part of the problem, people and technology must both also be parts of any *solution* to the problem. To help, people who produce new technical ideas need to know what to do, and to have some idea of whether these ideas are succeeding. Sustainability is now a field with few specific technical objectives and fewer means of measuring progress toward those objectives.

The word "sustainability" also suggests ideas that are (or at least, seem) deeply counter to those held by the cultures that have developed the most expensive habits. The developed world is now largely capitalist and not very successful in thinking about actions and reactions in terms of social return rather than financial return. As humans, we tend to try to solve problems through *action*: To build a better society, we mine coal and iron, and build cities, and grow our population, and try to make ourselves more comfortable. We often act selfishly. When faced with a problem whose solution may involve doing less rather than more – to conserve, to restrain, to shrink – and asked to act for a collective good rather than an individual good, we are not usually as effective.

Another issue with sustainability is that it is an idea that has evolved in developed societies, where there is generally enough to eat. It is attractive to those who have more than enough; it is substantially less interesting to the 80% of human kind who go to bed hungry and thirsty. To be told to eat less, and forego electricity, by those who have never thought of doing either is annoying.

So what could cause changes in behavior large enough to redirect a society from "consumption" to "sustainability" on a planetary scale? Catastrophe is usually the first specter to be trotted onto the stage. Climate change may be second, with its possible consequences: massive physical dislocations of societies by rising sea level, drought, and storm. Depletion of resources, and collapse of standard of living, is a third. War, large-scale movements of people, extinction of species, and so on – all are possible consequences of not acting sustainably. Although the consequences of business as usual are high, the consequences of ignoring sustainability all seem to be too distant to be real, it's inconvenient to change old habits, and we are asked to pay – in money, convenience, time, or comfort – for many changes.

Still, quietly, change does occur. Automobiles are an example. The automobile of today is safer, more efficient, and more comfortable than that of 50 years ago. Much of the improvement in automobiles has come as a result of government regulation – with the U.S. State of California often forcing global technological change – rather than consumer demand, but no matter. The air in cities is less polluted than it was, emissions have decreased, and we are much more likely to survive a crash now than even a decade ago. Unfortunately, in terms of sustainability, automobiles are also bigger than they were, and there are many more of them – since there are many more of us, and we all want the convenient personal transportation they bring.

Cars could, of course, easily be smaller and use less energy; the technology for much more sustainable personal transportation already exists. Since a large fraction – approximately one quarter – of energy use in developed societies is transportation, why aren't cars smaller, lighter, and more efficient, and why isn't public transportation more important? The answer is straightforward: We don't want them. Also, small cars are generally less profitable than large ones. Individual choice and capitalism have combined in a way that has made change in automobiles, and individual transportation, and petroleum consumption, and sustainability, slower than it might be if only technological readiness were important. In the developed world, we prefer the comfort and safety of big cars; in the developing world, we prefer cars to motorcycles and bicycles: "Bigger" and "more" are now clearly winners over "smaller" and "fewer." The uncertain future promised by sustainability is less attractive than the familiar present we know.

The story of the automobile has many interesting and instructive lessons for sustainability. The Internet has others. The Internet rests on a remarkable device: the transistor, or, if you prefer, networks of transistors stitched together into structures called integrated circuits. Transistors are just switches for electricity – not so different in function from the light switches on the wall – except that they have no moving parts, are unbelievably small, cost essentially nothing, and virtually never fail. From transistors, by a history no one imagined when they were invented, came computers, cell phones, the World Wide Web, the Internet, Google, and Facebook; also nuclear weapons, the evaporation of privacy, spam, and Facebook. Systems containing computers form the foundation of developed (and increasingly, developing) societies, in ways that no one could have predicted. Have they, collectively, contributed to sustainability or detracted from it? It is almost impossible to know although every instinct is that they will be absolutely essential to large-scale progress toward it in the future. They make travel less necessary, they create and destroy jobs, they are essential to every form of industrial activity, they consume enormous amounts of energy, they minimize the physical movement of information and maximize the amount of information being moved.

What is remarkable about both the automobile and the transistor is that they have changed society, but in ways that almost no one predicted. The world (or at least the most affluent parts of it) *is* more comfortable and convenient; we live longer and spend less of our life struggling for survival. We *do* use more energy and other resources to build that world. Can we have the benefits without the costs? Can we live happily with fewer benefits and lower costs? Are we willing to share the benefits with those who do not yet enjoy them and cannot pay for them? (Are *we* willing to pay for them?) All are interesting questions on the path to sustainability.

One of the characteristics of sustainability is that its costs and benefits are collective. If I buy a large car or a small one, I know what I'm buying. If I encourage change leading toward sustainability, I share the benefit, but may even have to pay in some way for someone else's benefit. There's a strong element of altruism and collective benefit in sustainability. "Altrusim" is not something that human societies carry out very well; also actions whose result is not quickly apparent.

Much of the discussion of sustainability is now focused on technological fixes, and often on fixes that provide more of something for less. Solar photovoltaics will capture sunlight and make electricity available without burning fossil fuel; wind generators will capture solar energy in a different way. The smart grid will make the transportation of this energy more efficient, and smart cities will allow it to be used more efficiently for lighting, and heating, and manufacturing, and transportation. It all sounds good, and I, personally, am for it. How much of a contribution it makes will ultimately depend on its cost, and what functions it enables, and how rapidly it can be deployed. But hypothesize, in some best case, new, sustainable, inexpensive sources of energy. Will they contribute to sustainability? If they reduce fossil fuel consumption and carbon dioxide production, yes; if they encourage population growth, probably no; if they accelerate the bloom of megacities, hard to know. Large-scale technological change, by itself, has always had unpredictable consequences. And large-scale technological change, including approaches ultimately intended for societal benefit, always requires large amounts of capital, and it is hard to keep capitalism, and profitability, separate from the discussion of societal benefit.

Regardless, and in any event, is engineering technology the key issue? Suppose that it is *social* engineering that really counts? Suppose that the popular opinion as to how the world should run is more important than technologies that might run it differently? It is possible that – up to a point – we will continue to prefer growth and consumption to alternatives with possibly brighter futures. One of the interesting changes is just beginning to emerge – a change enabled by technology, but influencing behavior rather than carbon dioxide levels – is social networking. The ability that will soon open to almost everyone to communicate (whatever that word may come to mean) with very large numbers of others – friends, enemies, and perfect strangers – through the Internet is something completely new – and may be a new normal. The democratization of communication may result (if flows of information can escape the organizations that might wish to control them) in types of collective action that are even now unimaginable. It is not clear what collective opinions large populations of people will evolve concerning sustainability or even if it will be (barring local disasters) a matter of any interest.

If, in the future, collective action is formed in the Internet, there is the possibility of influencing that action by manipulation of the Internet, for reasons both benign and malign. Universal connectivity is, however, the kind of large-scale change in capability that is almost certain to cause large-scale change in society. Sustainability may be an important part of this change, or it may just be a minor side show.

The ultimate argument supporting sustainability is that we can reach a better future for our grandchildren and great-grandchildren through responsible action now, and that wise, responsible, altruistic people acting appropriately probably would leave a better world. But suppose we are not wise, responsible, and altruistic? Suppose it all works out badly? It's useful to remember how adaptable people are. The circumstances of those who live in the slums in Kebare or Mexico City seem unendurably dire from the vantage of those living in an affluent American or European suburb, but these circumstances are those in which most humans now live, and in which almost everyone lived before industrialization changed everything. We are astonishingly adaptable.

Finally, we should remember that sustainability is still a baby idea, although one whose parents hope will grow up to become a big, important, and powerful one. But one needs to have more than aspiration – what sustainability needs is specificity, and new ways to engineer change and to change behavior.

George Whitesides

Contents

1   What Medical Equipment Taught Me About Sustainability   1
    Robert Malkin

2   Food: The Ultimate Answer   7
    José Andrés

3   A Village of Hope: The Interplay of Sustainability
    and Community Development   13
    Mark Templer

4   Trade-Offs in Sustainable Development   19
    Homi Kharas

5   Sustainable Skyscrapers and the Well-Being of the City   25
    Richard Cook

6   Dimensional Stability   31
    Heidi Williamson

7   Sustainability: A Tale of Twin Brothers   33
    Ken Wilson

8   Cooperation and Sustainability   39
    Simon Levin

9   The Sincerity of Purpose: Sustainability and World Peace   45
    Arun Gandhi

10	Recycling Reinvented: Music and Sustainability................................ José-Luis Novo	51
11	Connectivity and Sustainability: Perspectives from Landscape and Urban Design .... Diana Balmori	55
12	Nutrition and Sustainability.. Marc Van Ameringen	59
13	A Poet in the Car Company: Sustainability of Passion and Profitability... David Berdish	65
14	The Need for Sustainable Heretics... Freeman Dyson	71
15	Performance with Purpose .... Dave Haft	77
16	The Business of Sustainability: A Different Design Question.. Gregor Barnum	83
17	Sustaining Population Health .... Jacqueline Sherris	87
18	The Struggle to Make Sustainable Change in Global Health.. Laurie Garrett and Zoe Liberman	91
19	Economic Growth and Sustainability Rooted in Financial Literacy .... John Hope Bryant	95
20	Approaching the Future with Optimism... Robyn Beavers	101
21	A Decent Place to Live .... Jonathan Reckford	107

22	**From Field to Market: Changing Our Focus** .... Gerald Steiner	113
23	**Joules: The Currency of Sustainability** .... Chandrakant Patel	117
24	**Innovation Economics: The Race for Global Advantage** .... Robert Atkinson	123
25	**Unlocking the Energy of Business to Effect Change** .... Meg Crawford	127
26	**Put It on Paper: Lowering Healthcare Costs** .... Una Ryan	133
27	**Mind the Gap: A Different Take on Sustainability** .... Matthew Taylor	139
28	**Sustainable Scientific Research** .... Katepalli Sreenivasan	145
29	**Energizing Sustainable Development** .... V.S. Ramamurthy and Narendar Pani	151
30	**The Importance of Sustainability in Helping the Poor** .... Mechai Viravaidya	155
31	**Why Is Waste a Dirty Word?** .... Melanie Walker	159
32	**How Much Is Enough? Making It Personal** .... Toinette Lippe	163
33	**Teaching Sustainability in the Anthropocene Era** .... Kai Lee and Richard Howarth	167
34	**Don't Sustain; Advance** .... Kevin Finneran	173
35	**Changemakers for Sustainability** .... Karabi Acharya	177

36	What Social Entrepreneurs Taught Me About Sustainability	181
	Mirjam Schöning	
37	An Emotional Connection with Sustainability Through Documentary Films	187
	Heather MacAndrew and David Springbett	
38	Conserving Energy for Tomorrow	193
	Scott Tew	
39	The Sustainability of Ocean Resources	201
	James Barry	
40	Will It Last? Will It Endure?	207
	Andrea Coleman and Barry Coleman	
41	The Holistic Enchilada: Moving Toward Food System Sustainability	211
	Wayne Roberts	
42	Bringing Organizational Sustainability to Public Postsecondary Education	215
	Christopher Hayter and Robert Hayter	
43	Conservation Through Connections	221
	Harvey Locke	
44	Beyond the Status Quo: Catalyzing Sustainability in the Arts	227
	Jane Milosch	
45	Historic Preservation: The *Real* Sustainable Development	233
	Donovan Rypkema	
46	Bending Toward Justice: The Search for Sustainable Energy	239
	Michael Brune	
47	Afterword	243
	M.S. Swaminathan	

Chapter 1
What Medical Equipment Taught Me About Sustainability

Robert Malkin

The first time I stepped into an operating room in the developing world, I knew what I wanted to do for the rest of my life. It was a children's hospital in Managua, Nicaragua. I was travelling with a team of surgeons, nurses, intensivists, pump specialists, anesthesiologists and one engineer – me. It was the middle of a procedure to fix a heart defect in a beautiful little girl – a little girl with huge brown eyes and an innocent smile, despite a life filled with poverty and severe disability. She could hardly walk without shortness of breath due to her heart condition.

It was the middle of the surgery and it was intense. There was none of the usual banter among the surgeons and staff. This was a critical point in the surgery when the patient is "on pump," which means that her heart and lungs were bypassed and a machine was taking their place, controlling the little girl's temperature as well. Coincidentally, this was a machine I knew very well – a Sarns heart-lung machine. The software for the cooling unit on the successor to this machine was one of the first medical devices I ever designed and manufactured as an engineer.

What happened next was amazing. At a critical moment in the operation, the overhead surgery lights caught fire. There was smoke billowing from the lights!

Now, you would think that a fire in the operating room would cause quite a bit of panic. Indeed, there was a decent level of anxiety among the Americans – but not the Nicaraguan staff. They knew exactly what to do. They calmly took a small blanket that had been sitting on the side and placed it over the patient, then they called the technician on duty. The Nicaraguans clearly had seen this happen before.

The technician arrived, calmly removed the light bulbs from the smoldering fixture, and replaced them with regular 100-W light bulbs. All the while, the little girl was still on the table – and on the pump.

Meanwhile, I was trying to stop the technician. I told him that the correct light bulb had a rear heat reflector incorporated into the bulb. The heat reflector prevented the fixture from overheating – and, well, catching on fire! If he replaced the bulbs with regular 100-W bulbs, the fixture was going to catch fire again.

Without hesitating and without contempt or surprise, the technician told me that he knew that. However, the correct bulbs weren't available in Nicaragua. And, even if they were, they would be too expensive. He said, "It is surgery with the occasional fire, or no surgery at all. What can we do?"

> Sustainability means first and foremost that an outside effort can help launch a community towards achieving its objectives. Second, the outside effort should be designed such that the project does not continuously involve outside donations of resources.

At that moment, I knew what I needed to do for the rest of my life. Together with Dr. Mohammad Kiani and Cathy Peck, I founded Engineering World Health, or EWH, with the goal of helping to solve the problem at this hospital in Nicaragua.

But of course, it was not just this one hospital. The Director General of the World Health Organization recently said, "About 70% of the more complex devices [in the developing world] do not function ... only 10–30% of donated equipment ever becomes operational." The problem is worldwide.

Partially because the problem is so widespread, EWH has grown tremendously. In fact, EWH is now the world's largest provider of post-donation medical equipment service in the world. However, it was not always like this.

We started EWH with the idea of refurbishing medical equipment. It was a great idea. At that time, American hospitals were often retiring their equipment in fleets every few years. So there were literally tons of medical equipment available for donation. We would get just a bit of that equipment donated to EWH. Then we'd test and refurbish, and finally deliver the refurbished equipment to needy hospitals. In the meantime, the trips supporting short-term medical missions, like the surgery I described earlier, would continue as well.

And we did just that. We were shipping equipment all over the world: Africa, Central America, Asia – entire containers full of refurbished equipment.

Then a very strange thing happened. I was on a trip a few months after EWH had delivered a container of equipment to a nearby hospital. I had some time and I thought it would be great to see all our equipment making a difference. So I visited. But, I couldn't find the equipment. It wasn't in the OR or the ICU or the ER or the clinical lab. I did find a piece or two but the vast majority of it was nowhere to be found. I asked a colleague and doctor at the hospital what happened to our donation.

> Without sustainability, whatever outside effort is put into furthering the interests of a community leaves nothing behind when the outside effort ends.

He showed me something that I never expected to see. We drove across town to a large non-descript building. "We're renting this," he explained. Inside, I saw piles of medical equipment, a sea of boxes. Row after row of incubators, infusion pumps, ECG's, ESU's, medical supplies – everything. The building was gigantic and full of donated medical equipment, including most of the equipment EWH had donated. This hospital had received so much donated equipment that our donation was probably costing the hospital more to rent that warehouse than what it would have cost to buy the few donated pieces they were using on the used market.

Talk about not being sustainable, our donation had in fact *hurt* the hospital.

And this is not the only example. In 2003–2006, my lab conducted a survey of 54 hospital doctors and administrators in 16 developing world countries (see Malkin, R.: Annual Review of Biomedical Engineering 9, 567–587, 2007). One of their main complaints was that they had too much equipment. Not that they had too little. They had *too much*! We found warehouses of donated equipment in Central America, Africa, Asia, everywhere.

Now, just to be clear, the medical directors did not say that they had the right equipment. They had a lot of equipment they could not use. A lot of those warehouses contained pieces that were clearly broken. As the Director General stated, a lot donated equipment will never see operation.

At the root of this problem is that donations don't work. They are *not* a sustainable solution. Donating medical equipment fundamentally does not work to improve conditions in the developing world. In fact it sends the country backwards. And it is more than just a case of hospitals directors poorly managing a donation stream. Even when we donate a piece of medical equipment that is working and that meets the needs of the hospital, we still may be hurting the country.

Remember that the Director General said that 70% of some medical equipment in the developing world does not work. She was citing work that we did when I and a few others were appointed by the WHO to rewrite their medical equipment donation guidelines. That work was recently published in a larger form covering 112,000 pieces of medical equipment in 16 developing nations (see Perry, L., Malkin, R.: Medical & Biological Engineering & Computing 49, 719–722, 2011).

Consider this scenario for a moment. What if even half of the appliances in your kitchen didn't work? The toaster worked fine but the coffee maker wouldn't turn on. Your refrigerator worked but the stove wouldn't heat up. Could you cook? Do you think a doctor can practice medicine when half or more of the inventoried equipment is broken?

But there is a hidden, more insidious, finding from our recent publication. We found that about 5% of the medical was locally produced. The most common pieces of locally produced medical equipment were wheelchairs and lighting devices (billi lights, surgery lights, and exam lights). What is so striking about that figure? Those are also two of the most commonly donated categories of medical equipment. In other words, we are donating equipment that is being locally produced.

In fact, we just completed a survey of eight African medical device manufacturers. One of their main challenges was competing with donated, imported medical equipment. In essence, when you donate medical equipment, at least in these two common categories, someone in the developing world loses a job.

> *Don't donate.* Donating usually sends the community backwards. Rather, partner. Demand as much from the community – relative to their means – as you demand from yourself and your volunteers.

And, by the way, I am not just talking about donating *used* medical equipment. Among their complaints was donated *new* medical equipment. Much of that equipment is manufactured under contract in China. Many exciting technologies that promise to revolutionize healthcare in the developing world will eventually be manufactured under contract in China. Each one of these is a missed opportunity to create a new job in the developing world. Or worse, if the new device overlaps with locally manufactured equipment markets – like mobility and lighting devices – the new device might even put a local manufacturer out of business. That new, low cost design for the developing world could make things worse!

And, this is not just healthcare technology. This problem really deals with all aid (see for example, Dambisa Moyo's *Dead Aid*, 2009). All aid has the potential to hurt the recipients.

So what can be done? Conditions in developing world hospitals are dire. They need working medical equipment. We have it. But donating it is not a sustainable solution – it may not even a short-term solution.

First, ever since I saw that warehouse full of donated medical equipment, including equipment donated by EWH, we has refocused its efforts. We now donate almost no equipment. Yet we are one of the developing world's largest providers of working medical equipment.

The secret is that we fix all that equipment that other people donate! Through the EWH Summer Institute, currently managed by Duke University, we send over 50 engineers to struggling hospitals in Nicaragua, Honduras and Tanzania. Through a collaboration between the Edgerton Center at MIT and the Global Public Service Academies, another organization I founded, high schools students can also make a difference in developing world healthcare, on the ground, where action matters.

But, our students do more than just fix other people's donations. Students in my classes on Developing World Healthcare Technology at Duke University and students on the EWH Summer Institute devise alternative solutions to make equipment work. For example, students on the summer program have rewired dozens of surgery lights around the world to use the backup lights from trucks instead of expensive, custom light bulbs that aren't available in the country. They pull the old fixtures out, put in the truck fixtures. Now the surgery lights work and the bulbs can be locally replaced. And importantly, the rewired fixtures don't catch on fire!

And EWH does more for sustainable interventions. We also focus on capacity building. In 2010, I published a paper with one of my students analyzing 3,000 work orders and repair records from poor hospitals in 11 countries and 60 hospitals. From that analysis, we realized that there only about 115 basic skills that were required to repair over 60% of the broken equipment. We are talking about skills as basic as replacing a fuse, when you can't find an exact replacement in your market, or repairing a blood pressure cuff using a bicycle tube repair kit. This curriculum is now being taught in Rwanda, Honduras, Ghana and will soon be taught in Cambodia. The results from Rwanda are preliminary, but in the first matched study of the impact of training medical technicians, technicians trained using our new curriculum reduced the amount of out of service equipment in their hospitals by 35%.

Remember that little girl who was on the pump when the lights caught fire? She is fine. She recovered from surgery without any complications. In fact, she never knew about the extra excitement during the surgery or the remarkable change that occurred in my life during her surgery.

***

Robert Malkin *is Professor of the Practice of Biomedical Engineering and director of the Developing World Healthcare Technology Laboratory at Duke University. He is a founder of Engineering World Health. He received his Ph.D. from Duke University. Malkin serves as an expert advisor to the World Health Organization's Advisory Group on Healthcare Technology, Advisory Group on Innovative Technologies and serves on the World Health Organization's subcommittee on medical equipment donations. He is a fellow of the American Institute of Medical and Biological Engineering.*

Chapter 2
Food: The Ultimate Answer

José Andrés

In April 2010, I was sitting in Fond Verretes, a beautiful area in Haiti near the Dominican border. I was cooking a humble dish of old bread and canned sardines, with the help of a solar kitchen. Around me, watching me work with this solar cooker, were some locals and members of CESAL – a Spanish nongovernmental organization, with a few projects in Haiti and around the world. In a nearby area, children were playing with a ball made out of plastic bags that they had collected over time. I was struck by the expression of happiness on their faces. Looking at them, you would not imagine that their country just suffered a devastating loss of life, changing completely the way their nation would run itself over the next several decades.

For me, this was the latest leg of a long journey I began almost 20 years ago. In 1993, soon after I arrived in Washington, DC, I visited a nonprofit called DC Central Kitchen, located just minutes away from the U.S. Capitol. It was founded in 1988 by a visionary named Robert Egger, who was full of life and pragmatic ideas. With his incredible spirit, as well as a refrigerated truck and a small kitchen, he began doing something simple. He gathered up the untouched foods left over after the festivities of President Ronald Reagan's Inauguration. He brought them to the kitchen, where he repackaged them and sent them to homeless shelters and anywhere else people would be thankful for a good meal.

My life changed the moment I joined Egger's group as a volunteer. By watching Egger, I learned the true meaning of sustainability: what we give, what we may get, and how we share wealth with all.

The DC Central Kitchen's motto was straightforward: *Fighting Hunger. Creating Opportunity.* The organization grew into something bigger than a mere effort to feed people. A catering program, called Fresh Start, created jobs and income for the organization. The Campus Kitchens Project took the DC Central Kitchen's model and transported it across the nation to high schools, colleges, and universities. Ultimately, it helped me better appreciate the concept of sustainability.

> Sustainability is not the destination but the journey toward it. Every journey requires food. Creating hope through food can serve as an answer to some of our biggest challenges.

Above all, DC Central Kitchen created a culinary training program, taking people off the streets, helping to cleanse them of drug and alcohol abuse or the sense of not belonging, and training them to become cooks. In the process, they regained their dignity and became leaders of the many legions of volunteers who come to the kitchen every day. They would oversee regular people or CEOs and presidents of big companies. Those actions gave the students an extra layer of respect and self-worth. Nothing is more powerful and inspiring than to go to a DC Central Kitchen graduation, to hear their stories of achievement. I have shed many tears of sadness and joy over the years listening and learning from the men and women at the Kitchen.

I learned then that sustainability is not about the consumption of things: stuff, trees, mountains, or air. All of this has to be protected too. But the true meaning of sustainability is helping everyone in our community to become a rightful member, contributing to our society and to our future. Through people's pride, their right and desire to belong, food is the answer to some of our biggest challenges.

The knowledge I gained within DC Central Kitchen is what really compelled me to undertake my many trips to Haiti following the devastating 2010 earthquake. But the DC Central Kitchen model is one designed more for rich urban nations, with a few million living under the poverty level. The issues in a place like Haiti are of a different nature.

I'm a chef; I feed the few. But I have a great interest in learning to feed the world. To do that, I had to prepare myself. We have to listen and learn from the locals in the communities we want to help. So I founded World Central Kitchen with a simple idea: to help feed people in a sustainable way, using the model of a social business. Following the example of the microcredit pioneer and Nobel Peace Prize winner Muhammad Yunus, our approach is to invest in research and development, to try and stop throwing money at the problem, and really try to invest in the solutions.

> As a father, my responsibility is to help provide a good future for my children. In the absence of sustainable development, my children will not inherit a good society in which to live. One of the greatest hurdles to sustainable development – which affects food in a major way – is inequality.

So how can chefs really help the people of the world highly in need? In 1826, Jean Anthelme Brillat-Savarin, a great food philosopher, said, "The future of nations will depend on how they feed themselves."

But feeding the people of the world is not only about the food and how we process it. It's also about the hundreds of millions of people who don't have the right energy to cook those ingredients. And sometimes the "right energy," means having access to the tools that can provide sustainable and efficient cooking.

You and I only need a second or two to light up our stoves, or program the microwave. Seconds, just seconds. We blink our eyes and we start cooking. Millions of people don't have that opportunity. I've seen that firsthand.

Today a young girl in rural Haiti will be sent out to collect wood on the mountains because her family has no money for charcoal. Her life is put at risk. She will receive no proper education. Once back at home, her mother will make a meal using that wood in an inefficient cookstove, creating toxic fumes. Her mother, and maybe a young brother or two, will inhale those fumes. What is supposed to nourish them actually makes them sick and even is slowly killing them.

It only gets worse. Because the cookstoves burn wood or charcoal quickly, more time, effort, and money are needed that cannot be spent on education, food, buying seeds to plant, or improving their lives. With this incredible use of wood and charcoal, a country like Haiti is suffering widespread deforestation. Ninety-eight percent of the country has no trees, in a tropical climate. With no trees, there is no food. With no trees, there are no roots to create and sustain a healthy soil. When the rainy season arrives, which should become a moment of joy and celebration of the life that water brings, the rains don't quench the dry soil. Instead, it washes down the slopes of the mountains, creating erosion and often dangerous mudslides. Those waters become rivers, washing away the fertile soil, the home for seeds and food, washing away the homes of families, and sometimes taking more human lives. This is an environmental nightmare from the simple desire to cook food.

But there is real hope. A humble but powerful solution is to provide efficient, "clean" cookstoves to the people of the world. This challenging but attainable task will create many opportunities. The United Nations

Foundation, in partnership with the U.S. State Department, created the Global Alliance for Clean Cookstoves – where I serve as culinary ambassador – with the simple but powerful idea to bring 100 million clean cookstoves to 100 million households by the year 2020. When that happens, many girls won't have to climb a mountainside. They can receive an education and become a productive member of society. They will not grow sick from fumes. Their families will be able to buy books, clothing, and seeds to plant crops with all the money saved from charcoal or wood. With good soil to farm, life is possible all around. By selling crops from the farm or the fruit of the trees, these families are creating wealth, empowering their communities, all through the power of food.

Food is so much more powerful than we ever think. But we need to stop seeing food as a commodity. Eating food is the only thing, next to breathing, that we do from the moment we are born to almost the moment we pass away. Air and food are so alike. We need both to survive. If we do not take good care of food, we put ourselves and future generations at risk. Are the nutrients we get from the earth, and the way we get them we process them, good for our planet? If they are not good for the planet, they are not good for us. *Food, air, and water should be at the heart of our economic development and national security debate.*

> We need to appreciate the fact that sustainability is not just about economics or the environment. It's about smiles. It's about happiness. It's about the wealth of positive feelings within.

Obesity is a serious public health challenge in many developed and developing countries. People in poor neighborhoods have no access to fresh fruit and vegetables. The only way they can feed their families is with poor-quality, processed foods. Big retailers like Wal-Mart, Giant, or Safeway will not open in those neighborhoods. Why? Well, their business model is based on sales volume. I don't blame them.

But why don't we think about changing the model? We can invest in small fruit and vegetable stores all around the country by providing microcredit to a family that could own and operate a storefront or buy a small food truck. Business leaders can provide the training materials on how to start and be successful.

All of a sudden, a poor neighborhood has life, a corner with fruit and vegetables that will feed its people, with many of those products coming from local farmers. Those stores will take Supplemental Nutrition Assistance Program (SNAP) benefits provided by the government. With a simple idea we could create employment for many people. We could revive our inner cities and rural areas. We could give hope and self-esteem to people. We could fight obesity and other health issues, by providing better food choices and simple recipes that can be mastered by those who don't know how to cook. Local farmers will be able to connect with the urban communities they are supposed to serve, creating more jobs in rural parts of America.

We can take a government program offering food and multiply its effects 10-fold. A simple store in a poor neighborhood creates jobs, hope, wealth, healthier eating habits, safer neighborhoods, rural pride, and so much more.

Food is the only sustainable way to change the world. In so many ways, food is the solution.

***

José Andrés *is a culinary innovator, author, educator, television personality, and chef/owner of ThinkFoodGroup – a team responsible for renowned dining concepts in Washington, DC, Miami, Las Vegas, Los Angeles and Puerto Rico. His cookbooks include* Tapas: A Taste of Spain in America. *His native Spain awarded him the Order of Arts and Letters medallion in 2010 for his efforts to promote the culture of Spain, making him the first chef to receive this recognition. He founded the nonprofit World Central Kitchen and was recently named culinary ambassador to Global Alliance for Clean Cookstoves. He is Chairman Emeritus for the nonprofit DC Central Kitchen. Andres teaches "Science and Cooking" at Harvard, has been named an "Outstanding Chef" by the James Beard Foundation, and has been included as one of the "100 Most Influential People in the World" by* Time.

Chapter 3

A Village of Hope: The Interplay of Sustainability and Community Development

Mark Templer

I first set foot in India in 1985, visiting the city of Calcutta. I can still recall the thick, black smoke that hung over the city, the endless rows of bodies of people sleeping on the streets, dilapidated buildings and vehicles, and hawkers everywhere. I saw so many children, their tiny frames struggling to carry loads far too heavy for those so young. Yet in spite of the poverty all around us, we could see a vibrant, colorful, joyful society – proud of itself, struggling to survive, determined to make a better future.

One of my first days there I saw something that will be forever etched in my mind. On a narrow side street, there was a young boy, dressed in shorts, tied with a metal wire to a pipe on the wall. It was a disturbing sight. When I looked closer, I could see that he was blind. His parents must have been living on the street, far away from their village. They had gone to work, and there was no one to take care of him. So, for his own safety, they tied him to a wall until they could return later that day. It is easy to condemn others for their actions, but unless we try to understand their situations, we can never help them stand on their own two feet. His parents, among the poorest people in the world, had created a sustainable plan to build their family's future. It was not pretty, but it was realistic.

Other cities in India were less crowded than Calcutta, but their people were equally poor and had the same kind of noble courage and resourcefulness that I saw during my first days in India. Our small group of volunteers set foot in Calcutta, inspired by the examples of heroes such as Gandhi and Mother Teresa, determined to make a

> Sustainable development occurs when a community can continue to move forward economically and socially even if a partner ends its relationship with the community.

difference. Even if we could only help a few people in a sustainable way, we thought that it would be worth it. And so our adventures in South Asia began.

Over the years we built a charity (HOPE Foundation, the Indian affiliate of HOPE Worldwide), a network of volunteers and donors, and memories to last a lifetime. Although our work rarely got us invited to the 5-star world of NGO experts, it did genuinely change lives and help people to help themselves. Today that charity has over 500 employees in South Asia, educating 10,000 children on a daily basis, training 10,000 young people per year, caring for 1,000 orphans, and providing public health benefits to tens of thousands.

Like most charities, we searched hard for ways to stretch our money and make a lasting impact on the people we served. Our basic strategy was to find needy communities and make a long-term commitment to helping them to help themselves, starting with preschool education and simple health issues, and continuing with ongoing job training and literacy programs for young adults. Our staff worked especially hard at the grassroots level to build respect and true friendships with the people we served. They listened to both the outcasts and the leaders of the community and tried to shape our programming around felt needs rather than our own prescriptions.

One of the first communities we worked in was a leprosy colony in northeast Delhi with about 800 families. The colony lived in thatched huts with mud floors on land the government had donated for leprosy rehabilitation 35 years earlier. Over 90% of the residents sustained themselves *via* begging; despair filled the lanes of the Tahirpur Leprosy Complex. City officials were

> Sustainability is an attractive concept because it forces us to think about creating systems that mitigate risk and help us to be self-sufficient.

afraid to set foot in the colony due to the rage of the residents. At that time our charity had a corpus of funds and the good fortune to be working with Padma Venkatraman – the daughter of the then-President of India, Ramaswamy Venkataraman. Padma, typical of India's many amazing social workers, selflessly asked us to join her in helping the patients.

Through Padma's influence, the government of Delhi agreed to donate land, and our charity worked with the city to design and build a village for the leprosy patients – 800 homes in three phases. The city agreed to provide electricity, running water, roads, and sewage. We had countless meetings with the *pradhans* (community leaders) of the 17 leprosy societies there, and weekly meetings with the city, prodding them to keep their promises. Meanwhile we lobbied for the government to start a school there, and we started providing vocational training, preschool education, and microcredit. Twenty years later the colony, called the "Village of HOPE," stands completely transformed. Over 90% of the residents work. Young people are employed in call centers. A municipal school and government hospital have been built in the colony. People from neighboring communities mingle with the leprosy patients, who have built a strong network of local businesses. And for the past decade, the leprosy patients have donated regularly to others, starting with victims of the Gujarat earthquake in 2001 and the tsunami in 2004.

Through this program, and many others, we learned a few simple lessons about sustainable community development:

> Creating sustainable development in communities takes decades – not years. The expectation that a program or intervention can make a difference for a few years, suddenly withdraw support, and then expect "sustainable" results is just unrealistic.

1. The community's leaders need to influence the development of the program, and they should get credit for the program as it rolls out.
2. New leadership, especially involving local women, should be developed concurrently with program services.
3. Patient, positive engagement with local government officials is essential to ensure that problems are resolved constructively, and that citizens receive the services they deserve, without corruption.
4. Ideas that do not work should be discarded, no matter who thought of them.
5. Ongoing programs should directly improve citizens' ability to get education and jobs and to start new businesses.

Over the years many of our initiatives have not worked. Like many NGOs, at the beginning we trained people for jobs that did not exist or that did not pay (i.e., candle-making, chicken farming, or tailoring). We learned the importance of having full-time staff to network with local businesses to find out what skills they required and to place young people

> *What Worked*
> - Hiring job placement specialists to help trainees find work
> - Positive, rather than confrontational, engagement with government
> - Listening to and giving credit to local community leaders
> - Shifting from an output (how many people did we train?) to an outcome (how many people got jobs?) focus
> - Raising new leaders from the women in the community

> **What Did Not Work**
> - Running programs year after year without reviewing whether or not we still needed them
> - Training people in skills that did not prepare them to get jobs
> - Expecting donors to give year after year to the same programs
> - Keeping difficult employees longer than we should have
> - Trying to use broken-down used equipment (such as computers) instead of buying low-cost new equipment

in real jobs after their training. We learned the value of working with others – as long as you are willing to renounce credit and make other people look good, you can bring more resources into a community. We learned to remove staff who enjoyed arguing with others, and to subtly sideline difficult local leaders. We learned the importance of engaging donor partners in volunteering, so their employers and leaders could see firsthand the value of what we were doing. And we learned to bargain with the never-ending cycle of government managers supervising our programs – the transfer system ensured that we always had new people to deal with everywhere we worked.

Charities cannot replace government. But charities can form a powerful interface between communities and government, communicating local needs and holding government accountable to serve its citizens transparently and effectively. But until government changes the way it works, our work is only a drop in the bucket. Eventually, we have learned to work with city and state government leaders to recommend systemic changes in their health, education, and other social programs. Development that changes the way government works has a chance to be truly sustainable.

India today is very different from the one I first visited in 1985. The Right to Information Act is empowering citizens. The media is shining a light on corruption. Liberalization has allowed real economic competition and growth, providing tens of millions of jobs for people with skills. The charity work we did was especially effective because we were working in a growing economy. Without economic growth, job skills and even education might not lift people out of poverty. India's poor today still face incredible challenges. But the country is changing, and its people have hope. I am glad that our charity played a small, but real and sustained, part in making a difference to hundreds of thousands of India's very special people.

***

Mark Templer *has degrees in physics, economics, and political science from the Massachusetts Institute of Technology and the London School of Economics. After completing his studies, he worked overseas for 25 years, founding and serving as CEO of HOPE Foundation India, the South Asian affiliate of HOPE Worldwide. Templer recently began working for the U.S. government and continues to volunteer for the poor. The views expressed are his own and not necessarily those of the Department of State or the U.S. government.*

Chapter 4
Trade-Offs in Sustainable Development

Homi Kharas

I first went to Laos in 1987. As I drove the streets of Vientiane on my way to meet government officials, every child under 10 would turn to stare at my rented car, a rare and unusual sight on streets used primarily by bicyclists and bullock carts, with perhaps the odd motorcycle or rickshaw. Laos was just opening up to the rest of the world. Its communist leadership rarely met Westerners. It was a poor, landlocked, rural country with the highest concentration of unexploded ordinance on earth. Today it has joined the ranks of middle-income countries, thanks to the development of its hydropower and other minerals.

The example of Laos illustrates two points. First, economic development can happen fast, and when it does, the changes in people's lives are truly transformational. Second, at least in early stages, much economic development is not sustainable. Laos has cut forests, mined gold and other minerals, and dammed its rivers to generate electricity. These are the "low hanging fruits" of early development that generate resources and income, but mostly they are not sustainable economic activities. To achieve sustainable growth, Laos must invest in providing its people with education, health, and infrastructure and create a government that functions efficiently, effectively, and without corruption.

> When we have reached a point where our actions are no longer driven by narrow economic incentives but by our conscious decisions to do what will make us happier, we will have achieved sustainable development.

For most Laotians, sustainable development is an oxymoron. Development is what they need and want to achieve better, healthier lives free of poverty. However, it involves wrenching changes in society. Sustainability is about keeping to a traditional way of life, watching the Mekong slowly meander through town, and preserving the rich culture of the past. Laotians want both but probably cannot have both at the same time. For now, they have tilted toward economic growth and development, but at some point it is certain they will tilt the other way toward greater sustainability.

The beauty of the phrase "sustainable development" is that it encompasses two opposing concepts. Sustainability is about stability and maintaining the status quo, while development is concerned with establishing confidence in a better future and changes to achieve that. Surveys show that people value both stability and progress. Traditionally, an economist might insist on trying to find an optimal balance between the two, trading off sustainability against development at just the right pace. But that kind of technocratic approach does not square with how things work in practice. Policymakers and politicians resist trade-offs; they want to have their cake and eat it too. That is perhaps why everyone seems to be in favor of "sustainable development." It suggests that unpleasant trade-offs are not necessary.

But reality is harsher than rhetoric, and trade-offs may be inevitable. The 20th century will be remembered for many things. Its scientific and technical achievements, gleaming cities, and engineering marvels are unparalleled, but perhaps the most important achievement of the 20th century was the demonstration that billions of people can quickly enjoy far better living standards if they organize their national policies, institutions, and economies to take advantage of the opportunities that exist in the global economy. John Maynard Keynes already hinted at this in his essay on *The Economic Possibilities for Our Grandchildren* published in 1930, where he asserted that the economic problems of the advanced world may be solved within a 100 years.

The trouble is that we are finding that the most common path to progress is one that can only be taken at significant cost to the world's resources. Just at the moment, when the secrets to rapid economic development are being unraveled by more countries, it seems that the potential for development on a massive human scale could run across natural resource limits and planetary-wide boundaries. In this world, "sustainable development" raises fundamental questions about fair shares and access to resources across countries and across different income groups. What might be sustainable for the rich could condemn billions to poverty, and what might be development for the poor could condemn the planet to catastrophic climate change.

The fact is that today economic development is for everyone, not just for rich countries. Most recently, only 35 out of the 226 countries in the world were classified as low income by the World Bank. The rest have already achieved significant development and have gone through profound transformational changes. Although most countries subscribe to the doctrine of

> If we did not consider the concept of sustainability, the pressures to develop could easily take us down a dead-end path. Sustainability is appealing because it is the only way we can reach our destination.

"sustainable development," in practice the governments of poorer countries have been more concerned about development and less concerned about sustainability, while the opposite is true in rich countries.

For the time being, certainly at low levels of development, sustainable development does not seem to exist! Sustainable activities, like smallholder organic farming, do not seem to lead to development at the pace that many wish for. And many development activities, like logging, mining, and industrialization supported by energy subsidies, cannot be sustained for long periods of time. So it may be more appropriate to think of sustainable development as retaining flexibility to pursue either development or sustainability at different points in time, rather than in terms of pursuing both simultaneously.

Nobel Prize–winning economist Amartya Sen in his *Development as Freedom* describes development as a process of expanding freedom, in which one kind of freedom helps develop other kinds of freedom. But some development does have irreversible consequences and some freedoms may be lost along the way. Biodiversity is the obvious example. Once it is lost, it cannot be regained. Much the same can be true of culture. So one could argue that development has resulted in losses and reduced freedoms in some areas, even while expanding freedoms in other areas. On balance, this is probably positive and it is certainly true that richer, more developed societies try to preserve their cultural heritage and old traditions as well as their natural environments, so the tilt toward sustainability becomes greater as countries develop.

There are many examples of a cycle of destruction followed by preservation during the process of development. Social traditions can sometimes be sustained or expanded during development and help foster further development, or they can become a barrier to development. For example, the Indian caste system and some traditional Pacific Island customs around land ownership are traditions that may be incompatible with development and the expansion of freedoms. But once they are changed, secure property rights and mature social capital become significant underpinnings for development that need to be preserved.

> Taking one step backward may not be a problem if it positions us to take two steps forward. Understanding sustainability means recognizing we are moving toward a common destination.

Most people think of economic development as the process through which incomes are generated. But it is also about how incomes are spent. In fact, if development is about improving the quality of life and happiness, then at least an equal amount of attention should be devoted to the expenditure side of the ledger. Too often, economists fail to do this. Richard Easterlin focused attention on this issue with a set of surprising findings: Happiness, as reported in responses to surveys, does not appear to increase as a country's income level increases, and cross-country differences in happiness do not correlate well with cross-country differences in income levels (see the chapter entitled "Does Economic Growth Improve the Human Lot?" in *Nations and Households in Economic Growth: Essays in Honor of Moses Abramovitz*, 1974). Perhaps that is related to the fact that many of the things we enjoy doing, like spending time with family and friends, playing sports, reading, enjoying art, cultural events, and parks, are not measured in monetary terms and therefore do not fall into our metrics of economic development. But they should certainly be captured in the concept of sustainability because, at the end of the day, development should be about enabling us to lead happier, more fulfilling lives. This suggests that we should not label as "development" everything that leads to more material accumulation. We should instead understand development as leading to stronger social relationships and increased happiness. This point is well illustrated by the revolutions taking place today in the Arab world. Official aggregate statistics of economic development, like the growth of the gross domestic product (GDP), have been quite favorable over the past decade in many Arab countries, and social indicators (including girls' education) have improved faster than in any other region in the world. But even before the uprisings of 2011, Gallup poll results showed a majority of the population complaining about deterioration in their own situation. This gap between individual perceptions of their own circumstances and aggregate measures of development helps explain why 83% of the Egyptian population supported the overthrow of President Mubarak.

There is a prevailing fallacy that development is linear – a steady march toward progress that the sociologist Robert Nisbet traces back 3,000 years in Western civilization in his book *History of the Idea of Progress*. The very phrase "sustainable development" helps perpetuate this idea. After all, if development is sustainable (and sustained), then progress should be steady and continuous. This thinking is behind Western concepts of the

Enlightenment. If progress is driven by science, knowledge, and reason, then one can steadily build on what has come before – recall Sir Isaac Newton's famous remark, "If I have seen further it is by standing on the shoulders of giants." But there are other philosophies of development that question this. Some emphasize the role of chance, with negative outcomes almost as likely as positive ones. Others are more fatalistic and believe in divine destiny rather than in the idea of mankind in control of progress. Duality, or the theory of cycles of progress and reversals, also has proponents. In today's world, there is also considerable skepticism as to the ability of governments and policymakers to make rational decisions in support of long-term progress. So the notion that sustainable development is feasible is by no means universally accepted.

How do these ideas play out in the real world? I wonder if the gains that have been made in reducing poverty, educating children, and lowering child mortality in Laos will be permanent or transient. In the absence of effective political and social organizations, there is a high chance that there will be backsliding. More often than not, countries like Laos suffer from what economists call the "natural resource curse." The easy accumulation of wealth from mining or logging can become a tempting prize that encourages violence, conflict, and a willingness to do anything to remain in power.

In such circumstances, it is difficult to have a lot of confidence that the future will be better than the present. The risk for countries like Laos is that they may generate temporary development during a phase in which megaprojects generate growth and economic returns to a few project sponsors and the government, but that these resources are frittered away over time. Development is a long process – the advanced countries of the world have been on a positive economic development curve for about one and a half centuries, while most developing countries have only 30–50 years of development experience since their independence. Will they achieve sustainable development? One can certainly hope so, but in the immortal words of Chou En Lai when asked to comment on the historical impact of the French Revolution, it is probably "still too early to tell."

***

Homi Kharas *is a Senior Fellow at the Brookings Institution in Washington, DC. He has recently been a Non-Resident Fellow of the OECD Development Center, a member of the National Economic Advisory Council to the Malaysian Prime Minister, and a member of the Working Group for the Commission on Growth and Development. He has previously served as Chief Economist for the World Bank's East Asia and Pacific region. He holds a Ph.D. in economics from Harvard University.*

Chapter 5
Sustainable Skyscrapers and the Well-Being of the City

Richard Cook

In the summer of 2003, we had the remarkable blessing of being asked to design the most sustainable skyscraper in America. Even more compelling, the request came from the Durst Organization, a multigeneration family business that had previously developed the Condé Nast building at 4 Times Square – America's first green skyscraper. "Leave this place better than you found it," the philosophy of Jody Durst, has guided the Durst Organization through decades of sustainable building practices in New York City. Our task was to learn all we could from the first generation of green skyscrapers while pushing forward the standard in workplace performance for Bank of America, the primary tenant and joint venture partner in the building. It soon became clear that the design of a skyscraper in New York City was to speak powerfully about our generation's goals and aspirations. We needed more than a green skyscraper; we needed to set a new standard for urban sustainability.

My very first exposure to the practice of sustainability was in designing the Ross Institute in the late 1990s. The Ross Institute in East Hampton, New York, is focused on educating the whole person – mind, body, and spirit – and preparing students to become effective global citizens. The Institute's nontraditional curriculum trains students to think differently, balancing academic growth and personal

> As an architect of skyscrapers, I believe that the most sustainable buildings and cities are those that find their beauty and create a sense of well being through a connection with nature. Upon moving to the 49th floor of his new LEED Platinum skyscraper, one of our clients was asked to name his favorite "green feature" of the building. His response: "the view."

health with inner contemplation and community participation. As the centerpiece of this evolving campus, the Center for Well-Being is a 40,000-square-foot facility that focuses on themes of health and fitness by fusing Eastern and Western traditions. Through the design process, I developed a new-found focus on sustainability in order to address the challenge of creating a building meant to connect the user with nature via all the senses.

Soon after completing the Center for Well-Being, two sons came into my life, making personal the concept of global citizenship. We are an adoptive family, and our lives are forever tied to our twin boys' country of birth, Cambodia. My world view changed considerably by becoming a father; concern for the next generation became tangible. This exposure to the issues of global responsibility led to a personal conversion that shaped my focus on sustainability and the prevalence of the natural world as an imperative for my architectural practice back in United States. For me, practicing sustainable design stems from the *why*, before the *what* or the *how*.

Thankfully, we are witnessing a shift in public

> *Live Work Home*
>
> Grounded in ideas of healthy living and biophilia, the *Live Work Home* serves as a sustainable, social response to Syracuse's 21st-century concerns as a postindustrial American city. Our solution was an affordable, flexible space easily adapted for many household types as well as a home-based workshop and office, merging the often separated realms of "live" and "work." As we came to understand the Near West Side, plagued by high unemployment and blight, we realized the last thing the neighborhood needed was another single-family home. Instead, the area desperately needed work; its vitality is a question of sustaining livelihoods and social diversity.
>
> Our ideas about sustainable, long-term growth strategies for the Near West Side were influenced by the legend of the Three Sisters, a Haudenosaunee planting method in which corn, beans, and squash were considered "inseparable sisters" that thrived when interplanted. The legend reminds us that biodiversity and interdependence are essential to healthy human systems — like our natural ecosystem, successful communities are built from social and economic diversity
>
> The home is also a response to Syracuse's climate and ecology. Skylight tubes provide daylight for long, light-starved winters, and a perforated screen bounces daylight inside as if it is filtering through trees. An oversized front door opens to engage the sidewalk and street, creating an area of "prospect and refuge." To address stormwater issues, the functional landscape design includes bioswales

for on-site water retention and a green roof of low-cost, modular trays. A planted screen wall helps temper northwest winds, while native plants used throughout attract indigenous wildlife species. *Live Work Home* is a practical and replicable solution, and pursues sustainability as a core concept, achieved through thoughtful, common-sense solutions.

consciousness. Urban city dwellers are awakening to climate change and its alarming effects. This is a necessary refocusing, as the issue of gross consumption stretches much farther than just New York. Inefficient, outdated urbanity places an unsustainable demand on our sources of energy; therefore, it is imperative that we consider not only how much energy we use, but also where it is coming from and how it is made. PlaNYC also tells us that contemporary buildings will compose 85% of the built environment in 2030; the time to act is now.

The Bank of America Tower at One Bryant Park was designed to establish a new benchmark for high-performance buildings. It is the second-tallest building in New York City as well as the first LEED Platinum-certified commercial skyscraper. Instead of depending solely on the overtaxed energy grid of New York City, the building produces 65% of its own energy by using an on-site 4.6 MW cogeneration plant, which captures waste heat and lowers daytime "peak demand" by 30%. At night, when the building produces more energy than it can use, excess power freezes a group of ice storage tanks, which then melt during the day, providing cooling and further lowering peak demand. Through the use of waterless urinals, low-flow fixtures, and gray water treatment systems, we have achieved a 50% potable water use reduction, saving precious drinking water. Additionally, an element of the gray water systems is the implementation of storm water capture, which helps to alleviate New York City's serious problem of sewage overflow into rivers during heavy rainfall.

Furthermore, the design incorporates a new subway entrance, built just outside the front door, which links over 10 subway lines through a below-grade passageway to Times Square. Because these transit links allow efficient, low-impact commuting, 2.2 million square feet of real estate was developed without a single parking space on site. The Bank of America Tower not only sits above a major transit hub, but also reinforces the infrastructure of the city and increases the capacity of transit linkages. Accomplishments of this scale have radically reduced the impact on New York City's infrastructure and the environment.

In its namesake tower, the Bank of America occupies six high-tech trading floors and 75% of the interior. The design of these spaces has helped bring about a significant shift in the corporate real estate industry by acknowledging the higher value of a healthy, productive workplace

> We have witnessed a remarkable shift in the world population – from primarily rural to primarily urban – yet we all have an innate need to connect with nature. I believe the sustainability movement holds a key to global human health and well-being through its focus on our connection to natural systems.

for the thousands of employees who will spend their days in the building. To begin with, all energy-saving features in the building combined will reduce the annual energy costs by 50%, which totals over $3 million dollars of savings per year for tenants. As importantly, just a 1% rise in worker productivity due to the improved working conditions and indoor environmental quality will achieve $10 million per year in savings. To this end, we have incorporated a plethora of design features. For example, floor-to-ceiling windows and automatic light dimmers work in concert to saturate interior spaces with natural light and remarkable views to the outdoors. The inclusion of 300% more public space than required by zoning law blurs the conception of public and private space, and the Urban Garden Room on the ground floor provides an explicit connection to the natural elements of the neighboring Bryant Park.

Today, the built environment is beginning to reflect a growing awareness of the benefits of this connection to nature. This recognition can partly be attributed to the expanding field of biophilia: a study of the physiological and psychological effects of exposure to nature and a set of strategies by which we can weave nature into our built environments. Biologist E. O. Wilson described biophilia as humanity's innate response to nature and connection to natural systems. People understand on a subconscious level that a connection to nature feels good; to this point, it is the responsibility of designers to recognize the central role a connection to nature will play in the long-term efforts of practicing sustainability.

As much of an impact as the Bank of America Tower has made, it is just one building. Sustainable design must answer not only to advances of resource conservation or emission reduction but also to whole systems

> We need a much clearer understanding of the impact our infrastructure has on larger systems, specifically how energy is produced. In 2004, the electricity grid in New York City provided only 27% efficient energy. By creating power on-site at One Bryant Park, we increased the production efficiency of energy to 77%, reducing stress on an overtaxed infrastructure and setting a standard for marketplace transformation.

thinking. As we move forward in our efforts toward environmental responsibility, we must remember the fundamental need to connect with nature, especially in a widely urbanized world. By reconnecting with the natural world and ecosystems that support life, we can take the lead in the transition to a sustainable future, a movement that holds a key to global human health and well-being.

***

Richard Cook *is a founding Partner of Cook + Fox Architects and Terrapin Bright Green, both based in New York City. Over the past 25 years, he has built a reputation for innovative, award-winning architectural design. Cook + Fox has been recognized as designers for the Bank of America Tower in New York City, the world's first LEED Platinum commercial skyscraper. Cook's work has been showcased at the National Building Museum and in feature programs on PBS, the Discovery Channel, and National Geographic.*

Chapter 6
Dimensional Stability

Heidi Williamson

The way, for example, paper
retains its self despite
damp or heat or frost,
or a bird continues to flit
on altered airstreams.

*

Out here in the damp margins
flint fractures the fens' panorama
with bright towers of churches,
primed to call out if the sea
visits hastily, with no warning.

the sites of special scientific interest:
the heath of silver-studded blues;
woodlands where the pool frogs spawn;
lagoons where starlet sea anemones
practise their minute fission.

Instead, the sea deftly extends
its lips to the land, worrying
stones into flecks of sand:
all substance carried under,
like the story of the lost town.

Submerged into history,
the lungs of its chapel replete

with creatures, the stone floor
of its aeons-old sea burgeoning
beneath the new town, unsettled.

Even now I hear the *creak-crack*
of the pill boxes, rusting their way
towards France, their iron bars released,
as the houses and people and matter
progress and regress.

*

The way, for example, the ocean,
freeing itself of its borders,
falls constantly into its element:
its heedless waters closing
over our mouths.

> "Sustainability" feels like an intangible word when, in fact, it is incredibly tangible – every area is impacted by it. More clarity and visibility promoting all the work going on across many different arenas can help make it a real part of daily life. We're a part of the world, not apart from it, and practicing sustainability enhances our relationship with it – and ourselves.

***

Heidi Williamson *is a U.K.-based poet with an interest in science. Recently, she was poet-in-residence at the London Science Museum's Dana Centre. Her first collection,* Electric Shadow, *is a Poetry Book Society Recommendation and was short-listed for the 2012 Seamus Heaney Centre Prize for Poetry. "Dimensional Stability" will appear in Heidi's upcoming collection, to be published by Bloodaxe Books.*

Chapter 7
Sustainability: A Tale of Twin Brothers

Ken Wilson

"Crazy tree huggers." That's the kind of slur we evangelicals have thrown at environmentalists. Yet the founder of Christianity himself explicitly forbids contemptuous name-calling.[1]

Wouldn't it be powerful if evangelical leaders apologized directly to environmentalists for such behavior? After all, we view creation as a gift from God that we are called to steward. Environmentalists should be our partners – not our enemies.

My evangelical pastor colleague Tri Robinson did just that a conference of Idaho environmentalists. His apology opened up a wonderful dialogue about the positive role faith can play in fueling the environmental movement.

But the rift between those who might join forces to care for planet earth runs deep. Even twin brothers can find themselves on opposite sides of this cultural divide.

I first met Charles McNeill at an ambitious conference in 2009. Representatives of world religions were being paired with environmental leaders for a few days of dialogue. Charles is a biodiversity expert and senior policy advisor on environmental issues from the United Nations Development Program.

During an evening get-together, Charles asked me what kind of church I'm affiliated with. "Vineyard churches," I replied. Vineyard is an evangelical

---

[1] "You have heard that it was said to those of old, 'You shall not murder; and whoever murders will be liable to judgment.' But I say to you that .... whoever insults his brother will be liable to the council; and whoever says, 'You fool!' will be liable to the hell of fire" (*Matthew* 5: 21–22, English Standard Version).

Christian denomination with over 1,500 affiliated churches worldwide.

Charles' eyes widened, "My twin brother attends a Vineyard church in Los Angeles!" It turns out to have been one of the very first Vineyard churches.

> Sustainability is an essential criterion of morality: the application of the golden rule to the unseen others who are yet to come.

Later, my fellow Vineyard pastor Tri Robinson, who attended the conference with me, told us about apologizing to the gathering of Idaho environmentalists for the way evangelicals have regarded them. I couldn't help but say to Charles, "Wouldn't it be powerful if we could get the leaders of the American Evangelical movement and the American Environmental movement together in a room to apologize *to each other*?"

Charles' eyes turned quizzical, "What would the environmentalists apologize to the evangelicals for?"

What indeed?

The "Jesus Movement," a religious revival of the 1970s, powerfully influenced the current leaders of American Evangelicalism. The Jesus Movement came before the emergence of the Religious Right – in its early days, the Jesus Movement was antiwar, antiestablishment, and pro-ecology. In fact, the Jesus Movement was set up to be a powerful voice for environmental stewardship – a destiny it has yet to realize.

The Jesus Movement was absorbed into the broader American Evangelical movement, which, by the 1980s, had become powerfully aligned with a conservative political agenda. Sadly, conservatives have generally viewed environmentalism with suspicion. The environmental movement, in turn, has been shaped by a strong secular sensibility. Rather than viewing religion as a natural ally, many environmentalists viewed religion (and in America, religion often means faith shaped by the Bible) as a root cause of environmental abuse. These two movements – cut from the same cultural cloth in their early days – parted ways.

Now, like the two brothers in Jesus' famous story about a man with two sons,[2] these two movements – evangelicalism and environmentalism – find themselves at odds.

---

[2] Jesus' longest parable is the story of a Father with two sons; the younger brother returns to the Father's house after spending his inheritance in a far-off country. Scandalized by the Father's welcome of his younger brother, the older brother stays away from the party thrown for his brother by the Father. The story ends with the Father imploring the older brother to come back into the house to join his younger brother and the other party-goers (see *Luke* 15: 11–32).

> Sustainability appeals to my identity as an image-of-God bearer equipped with a capacity to care about a future that does not pertain to my immediate self-interest.

As a result, we've been unable to muster the cultural and political will required to deal with a slow-motion, but inexorable, disaster like climate change. *But why should this be?* After all, the sacred text of Evangelicals views our Father's house – the realm of nature – as a gift that should be treated with wisdom and respect.

As I spun the tale of how the early Jesus Movement parted ways with the early environmental movement – like the two brothers in Jesus' parable – Charles' eyes filled with tears. "Ken, you are describing me and my brother. While he was getting involved in the Jesus Movement in the 1970s, I was studying ecology at Berkeley, and I *did* learn to view religion as part of the problem, not part of the solution."

My own awakening to the importance of environmentalism took place at an earlier meeting of American evangelicals and environmentalists held at a secret location in 2006. James Gustave (Gus) Speth, the first advisor to a U.S. President on global warming, addressed the retreat participants, saying:

> Thirty years ago, I thought the top three global environmental problems were biodiversity loss, ecosystem collapse, and climate change. I was convinced that with enough good science, we would be able to solve these problems. But I was wrong. The real problems are bigger than that. They are things like selfishness, greed, and apathy. For those kinds of problems, good science isn't enough. For that we need a spiritual and cultural transformation. And we scientist don't know how to do that.

Then looking at the evangelical leaders at the table, Speth said, "We need your help." The hair on my arms stood on end. My throat tightened, and my eyes filled with tears. I was having a spiritual awakening to environmental concern. That moment was a powerful turning point for me. I wanted to be part of the solution, not part of the problem.

At the same retreat, renowned Harvard biologist E. O. Wilson took a bold step. He urged his fellow scientists, many of whom were secular in orientation, to reach out to the evangelical community as potential allies in the effort to address our global environmental crisis. Wilson himself had grown up in the evangelical world, although he eventually became a secular humanist. The culture wars of the past 30 years were fought between people Wilson knows and loves. At the retreat, Wilson suggested that environmental scientists begin the process of reaching out to people of faith by referring to the natural environment with its religious name, "Creation" (see E. O. Wilson's *Creation*, 2006). Wilson knows that we can't hope to work together until we learn to speak each other's native tongue.

As I began to interact with environmental scientists, I was surprised to see how similar they were in outlook to fellow evangelicals. Both groups approach the world with missionary zeal: We see ourselves as the bearers of news that the world desperately needs to hear; we view ourselves as a minority voice largely ignored by the majority population; and we share a pent-up frustration consistent with this shared outlook.

It is because of these commonalities that, despite the carnage of the culture wars, I'm hopeful that evangelicals and environmentalists can learn to speak to each other again, can learn to love each other again, and can learn to work together again as brothers and sisters.

It simply requires us to be true to our primary narratives. The primary narrative, the story that shapes the evangelical outlook, is a love story of God in search of humanity. It is a story that places humans as image bearers of God on the earth, set here to rule with the benevolence of a loving creator in love with his creation. We are here as stewards of a sacred trust, accountable to God for the condition of the trust passed on to future generations. We evangelicals have a great deal of work to do in order to be true to our primary narrative. We cannot be faithful to our founder without becoming better stewards of the natural world.

We need you (the sort of people who read books on sustainability) to help us (the sort of people who don't) in order to become better evangelicals.

> To unabashedly integrate our understanding and practice of sustainability with our deepest yearnings, expressed in of our irrepressibly religious nature.

The primary narrative of environmental science is evolution: the gradual emergence of life in all its glorious diversity from common ancestry by a process of adaptation – taking what has survived the test of time and nature's selection and adapting it to new challenges. Rather than create a new "environmental ethic" from scratch, the wisdom of the primary narrative of modern science would suggest that we look for what's best in the ethic of the surviving world religions and build from that. Perhaps along with new words like "sustainability," we could also use old words like "stewardship" – receiving what has been given in trust and passing it on to future generations in equal or better condition. You need us to help you become better environmentalists.

Charles and his brother need each other to care for their father's house. Let us begin by learning each other's native language in order to communicate across this infernal cultural divide. A matter of great consequence – the sustainability of human life on planet earth – depends on it.

***

Ken Wilson *is the Senior Pastor of Vineyard Church of Ann Arbor and the author of* Jesus Brand Spirituality: He Wants His Religion Back *and most recently,* Mystically Wired: Exploring New Realms in Prayer. *Wilson is a leader in a nascent movement to awaken evangelicals to environmental stewardship, serving on the Creation Care Advisory Group Task Force of the National Association of Evangelicals. Wilson is also the co-founder of the Friendship Collaborative, bringing evangelicals together with environmental scientists to explore common ground.*

Chapter 8
Cooperation and Sustainability

Simon Levin

I was trained as a mathematician but driven to become an ecologist in large part because of a concern for the future of my children and (then hoped-for) grandchildren, and because I was convinced that population growth and the insults humans were heaping upon our environment threatened their future. I have never believed, however, that the protection of the quality of life for humanity and of the services, tangible and otherwise, that we derive from natural systems involves a zero-sum game – between ecological protection and economic growth. Indeed, we have seen since the recent financial crisis that economic contraction diminishes our ability to take the steps necessary to deal with global environmental issues such as climate change and biodiversity loss. It also further exacerbates the problems of intragenerational inequity, poverty, and disease spread and fosters conflict among peoples that diverts us from our joint task of achieving a sustainable future. Cooperation is thus essential, both in how we manage our environment and in how different disciplines and different sectors view the balance between economic growth and environmental protection.

> Sustainable development means achieving the goal enunciated by the Brundtland Commission – to ensure that future generations are presented the same choices with regard to their quality of life as we have.

What does sustainability mean to me? It means achieving the goal enunciated by the Brundtland Commission (*Our Common Future*, 1987), to ensure that future generations are presented the same choices with regard to their quality of life as we have. It appeals to me as the fairest criterion for what we should demand of our environment, and my prescription for improving how sustainability must be addressed is that it needs to become more interdisciplinary, building partnerships across disciplines that at best have been independent of each other, and at worst have been at odds. The challenge of sustainability is not one for natural scientists alone, not one solely for physical scientists, and not one just for social scientists. It involves the interplay between ecosystems and climate, and between these coupled environmental systems and socioeconomic systems; but it also involves matters of ethics and fairness, and how we deal with the challenges of intragenerational and intergenerational equity.

Addressing the challenges of sustainability thus must rely on cooperation, both in logistics and in focus. First of all, we must build partnerships, across disciplines and across sectors, from ecologists and molecular biologists to physical scientists and climate modelers to economists and sociologists and psychologists; from scientists to ethicists; and from academia to the corporate sector to government and citizenry. Interdisciplinarity has never been easy to achieve in most branches of science: Scientists are judged, for good reason, by how proficient they are within disciplines. This is then reflected in their training, which is usually narrowly focused on the fundamentals of their subjects, and reinforced by the structure of universities and the tenure system. Most graduate students, and indeed most faculty, rarely venture outside the buildings in which their departments are housed, and indeed may be similarly constrained within subdisciplines.

We need to find ways to foster interdisciplinarity. How do we do that? The great mathematician Mark Kac once wrote that "If a mathematician wants to make a contribution" in "discovering the laws of nature," the mathematician must "in effect, cease to be a mathematician." Only then does play become science. "Perhaps it is well to be reminded, by way of analogy, that while in recent years a number of physicists have made significant contributions to biology, they accomplished this not because they were *physicists*, but because they became biologists" ("On applying mathematics," *Quarterly of Applied Mathematics*, 1972). Of course, in this way, new disciplines are born, but multicellular ones made up of pieces from previous evolutions; this imitates how the great transitions to multicellularity and cooperation took place in the evolution of the biosphere and is a model for work on sustainability. We do indeed need a new, multicellular science of sustainability, built firmly on the foundations of better-developed disciplines, and whose practitioners are multilingual. I have been fortunate to be involved in a few efforts to foster such interdisciplinary marriages, often met with the sorts of resistance within the disciplines that characterize intergroup marriages in other spheres. They have,

> Sustainability appeals to me as the fairest criterion for what we should demand of our environment.

however, been the most rewarding experiences of my career and I think the best such enterprises for building the foundations of the needed new science.

Achieving cooperation among different players in developing the science perhaps can serve as a model for the real challenges, finding cooperative solutions among individuals to build societies for now and the future, and among nations to ensure a sustainable future. We live in a global commons, in which the actions of individual agents affect others, and yet in which the incentives for prudent action are not sufficient to assure sustainability. We all draw resources from a common pool and exude wastes into each other's environments. Yet we all see our actions, correctly, as having little impact on the broader environment, and hence we are not sufficiently motivated to take serious action to restrain our own selfish tendencies. I am asked by hotels to reuse my towels to save the world, but it is hard for me to accept that this will do the job, or even that the hotels really have other than their own self-interest in mind in asking this of me. We need meaningful steps toward sustainability that engender trust, not token efforts that at best seem to be drops in the bucket.

William Forster Lloyd highlighted this problem nearly two centuries ago in his *Two Lectures on the Checks to Population* (1833), proposing what ecologist Garrett Hardin later termed the "Tragedy of the Commons." Hardin crystallized the issues sharply and said that the solution to the problem was "mutual coercion, mutually agreed upon." (*Science*, 1968). Indeed, a half-century after Hardin's seminal paper, the problem remains: How do we achieve cooperation in societies, and indeed in the global society, to address the issues that most threaten a sustainable future, be those issues economic, environmental, or simply how to coexist with one another? What can we learn from how evolution has dealt with these questions over billions of years?

Why cooperation exists, including especially extreme forms like altruism and eusociality, have bedeviled evolutionary theory since Darwin; indeed, Darwin delayed publication of *The Origin of Species* for 20 years while wrestling with what he regarded as a challenge to his theory of natural selection – why individuals give up their own apparent fitness to aid others.? The most extreme examples of this were primarily to be found in the haplo-diploid insects, in which males are haploid (carry single copies of each gene, from their mothers), while females are diploid (have two copies, one from each parent). Under such genetics, full sisters share three fourths of their genes, and thus relatedness is high. Furthermore, since even among fully diploid organisms it is apparent that individuals are more likely to aid their close relatives, the simplest explanations of high degrees of cooperation were based on close genetic relatedness, for example, in the work of the late evolutionary biologist W. D. Hamilton. But relatedness is clearly not the whole story, and indeed its importance remains a hot topic of debate. Individuals form reciprocal arrangements with others with whom they can expect to interact frequently, and nations do the same with other nations.

The late Nobel Prize–winning Elinor Ostrom led the way in illustrating how such reciprocal arrangements can make some groups effective in dealing with limited resources (*Governing the Commons*, Cambridge University Press, 1990). More generally, numerous examples of cooperative arrangements that provide mutual insurance in a fluctuating environment can be cited in human societies, like herdsmen, as well as nonhuman groups, including things as simple as bacteria and slime molds. Within human societies, customs and norms help enforce behaviors that increase the collective good, and these eventually may become formalized in rules and laws. The glue that holds these arrangements together is prosociality, in which individuals act in ways that, on the surface at least, show concern for the welfare of others. Prosociality clearly is encouraged by social norms, involving reinforcement mechanisms from punishment and taxes to approbation and reputation. Such mechanisms clearly exist not only within cultures, but also have achieved some genetic basis, leading individuals to behave in ways that seem to violate the simplest rationalist assumptions of profit maximization, most simply because individual utility functions are influenced by pride in one's actions.

> I am asked by hotels to reuse my towels to save the world, but it is hard for me to accept that this will do the job, or even that the hotels really have other than their own self-interest in mind in asking this of me. We need meaningful steps toward sustainability that engender trust, not token efforts that at best seem to be drops in the bucket.

To achieve sustainability, we must harness this prosociality, both toward contemporaries and toward future generations, and find ways to enhance it. One of the hopeful lessons from evolutionary theory is that cooperation is possible, at multiple levels, and that cooperative associations can evolve to become multicellular entities with a common purpose. No invisible hand, however, guarantees that these entities will serve the interests of all equally; hence, we will need new organizations, new associations, and new institutions for governance that foster the "mutual coercion, mutually agreed upon" that Hardin prescribed. Observations of current political systems and of current trends in international relations perhaps do not foster optimism that we can achieve that goal, but we have no choice. We are all in this together, and the only positive aspect of the acceleration of environmental damage is the hope that it will lead us to discount the future less steeply. In particular, we clearly must gain a better understanding of human behaviors, and of the factors that enhance other-regarding behaviors. That has led Stanford's Paul Ehrlich and others to create the Millennium Alliance for Humanity and the Biosphere, a first step in a long journey. We need other such organizations.

One of the most important experiences in my career has been the ongoing dialogue between ecologists and economists fostered by the Beijer Institute of Ecological Economics, in Stockholm, Sweden. The Beijer has built a conversation, a community, over more than two decades, bringing leaders from the two disciplines together, enhanced by others from complementary disciplines from game theory to ethics, and showed that it is possible to find a common language and a common purpose in addressing issues like the carrying capacity of the environment, the robustness and resilience of ecological and socioeconomic systems, and whether we in individual nations are consuming too much to permit a sustainable future. The Beijer group has developed metrics for sustainability and made the first efforts to quantify them, integrating the diverse perspectives of scientists and scholars from multiple perspectives. My hope is that it will serve as a model for other such efforts – not advocacy groups that advance a particular perspective, not even forums for compromise, but rather meeting grounds where all in the end are convinced they have found the right solutions. This is what my own experience in the Beijer group has meant to me; it is cooperation in the extreme, and has provided me with hope that we can find the pathways to sustainability that originally inspired my career switch.

***

Simon Levin *is George M. Moffett Professor of Biology and director of the Center for BioComplexity at Princeton University. His research interests are in understanding how macroscopic patterns and processes are maintained at the level of ecosystems and the biosphere, in terms of ecological and evolutionary mechanisms that operate primarily at the level of organisms; in infectious diseases; and in the interface between basic and applied ecology. Levin has been elected to the American Academy of Arts and Sciences, National Academy of Sciences, American Philosophical Society, and Istituto Veneto. He is a recipient of the Kyoto Prize in Basic Sciences, Heineken Prize for Environmental Sciences, Margalef Prize in Ecology, MacArthur Award in Ecology, and several honorary doctorates.*

Acknowledgment The grant from the National Science Foundation (DMS 0955699) to Princeton University is gratefully acknowledged.

Chapter 9

The Sincerity of Purpose: Sustainability and World Peace

Arun Gandhi

To understand what my grandfather, Mohandas K. Gandhi, meant by *sustainable development*, you need to understand the philosophy of his life. Gandhi's very essence was based on the concept of *Satyagraha* – that is, *truth force*. For Gandhi, truth was God. Thus, Gandhi's life involved not just uttering the truth at all times, but living, believing, thinking, and doing nothing but the truth. Gandhi also believed that service of the oppressed was the best form of worship and that poverty is the worst form of violence. In Gandhi's mind, as long as poverty existed in society, everything else was meaningless. He believed you cannot talk to the poor about freedom, spirituality, morality, or anything else so long as the poor are deprived of the three essentials of life – food, clothing, and shelter.

> We first need to understand what we wish to sustain. In one important sense, sustainability is linked with sincerity of purpose. If we know what we wish to sustain and if it is a righteous cause, then through sincerity we can sustain it as Gandhi did with his belief in the goodness of all human beings.

Gandhi often faced a conflict with the other leaders of the Indian freedom movement, who could not understand why Gandhi wasted time, energy, and resources addressing issues such as poverty, untouchability, emancipation of women, and education. After all, they thought the most important task was simply to free India of British Imperialism. All these other problems could be taken care of after independence, they argued. Gandhi's response? *Freedom will be meaningless to those who are starving, destitute, and ignorant.*

When a social worker asked Gandhi for a talisman, Gandhi wrote:

Whenever you are in doubt, or when the Self becomes too much with you, apply the following test: Recall the face of the poorest and the weakest person you may have seen and ask yourself if the step you contemplate is going to be of any use to that person. Will that person gain anything by it? Will it restore to that person control over his life and destiny? In other words, will it lead to *Swaraj* [self-rule] for the hungry and spiritually starving millions? Then you will find your doubts and yourself melting away.

Clearly, the phrase "sustainable development" standing on its own would mean different things to different people. Ask a hungry person, and he will say it means any development that will put food on his plates; a clergy would look at it from the perspective of spiritual and moral growth; a businessperson from the profit motive, and so on. Context, therefore, is absolutely essential. Indeed, for the poor and destitute, *sustainable development* has to initially mean *economic development*. I feel certain that for Gandhi the phrase "sustainable development" meant a form of economic development that would, first, sustain the lives of the poor in the Indian villages and, second, be of the kind that could be sustained by the economy of the country for a prolonged period of time.

Gandhi's mission was to gain political and economic independence for India. In this context, he often spoke of creating an Indian *Ram Rajya* – kingdom of Lord Rama. After all, Lord Rama was someone known to be extremely compassionate, humble, inclusive – and someone to whom truth meant God. Unfortunately, this led to misunderstanding, because it implied that Gandhi held visions of a Hindu and not a secular India. This was clearly not the case.

According to the story, Lord Rama became king of the nation-state of Ayodhya in northern India, where the people lived in utter happiness and contentment. The king was just and accessible to anyone at any time; the kingdom made no distinctions of any kind not only between humans, but even between animals and all living creatures. Rama was honest to the core and was known to disguise himself and walk incognito among the populace to discover whether there were any latent grievances or discontentment. If he detected any, he worked on addressing them immediately.

Thus, when Gandhi expressed his view that India should emulate the historical *Ram Rajya*, he didn't mean it in the religious sense, but rather in the moral sense. Gandhi's writing needs to be read with an open mind and in the context of his comprehensive philosophy of life. Except for his incomplete autobiography, Gandhi never wrote a book. Instead, he wrote thousands of articles and spoke at length on many issues. These were later collated by editors to form a number of books focusing on various issues.

The Sincerity of Purpose: Sustainability and World Peace

Misunderstandings and misinterpretations are the bane of human society, but sometimes both opposing meanings are equally valid. A case in point is the current controversy over hydro-fracking in the U.S. To some it means the destruction of the environment and the natural resources of the country, while to others it means the creation of more jobs at a time when millions are unemployed. And yet another group looks upon hydro-fracking as "sustainable development" resulting in a boost to the economy of the country and profits for the industrialists. Who can be an impartial judge in a contentious issue like this?

> Growth is a natural progression of life, whereas development (in sustainable development) is *planned* and implemented by human ingenuity to meet the needs of the people. To put it differently, a forest grows naturally in an unplanned and haphazard state, whereas agriculture is developed by human beings with proper planning to meet the needs of the market.

The concept of "development" in sustainable development, in my view, does not mean a constant change but a *planned change,* whereas "sustainability" could assume different meanings depending on the context in which it is used. Ultimately, what the two words mean to any individual would depend upon the individual's understanding and intelligence.

An example is when Gandhi told two people – one man and one woman – to go out into the most remote villages of India and serve the needs of the people. The man went and found a remote village that was steeped in poverty and ignorance. The people were unclean, had very little understanding of hygiene, and lived in squalor. He found it difficult to live among them, so he drove every day to the outskirts of the village and called the people to assemble under a tree far from their village. Here, he lectured to them about cleanliness and hygiene. The villagers heard his lectures but continued to live as they had always lived. Several months later, the man came back to Gandhi and complained that the people in the village are incorrigible. After all, he had spent 6 months trying to teach them cleanliness, but it had made no impact on them. Gandhi asked, "Did you find out if they had a water supply?" "No," said the man. Gandhi then asked, "If they do not have water, how do you expect them to be clean?"

On the other hand, the woman found a remote village with the same problems. She decided to live among them, thus building a rapport with the women of the village and acquainting herself with the villagers' problems. She got help from the nearest town to locate an underground water supply so that a bore well could be dug. This saved the women from walking 10 miles to carry water on their heads every day. The abundant water supply in the village changed their lifestyle. Once they had water, then cleanliness and hygiene made more sense to the people – new customs were quickly adopted. The woman addressed every little problem directly, often getting the people to find the solutions themselves.

> Sustainability is very directly linked with world peace; the reason why we haven't been able to achieve world peace is because we don't know what we want. We also don't know how we want to achieve it, so we are pursuing half-hearted approaches. Such approaches cannot be sustained, and we have therefore resorted to sporadic attempts.

If Gandhi was looking at the present challenges of Indian democracy, he'd be very distressed – just as he was when he and his philosophy were abandoned by Indian leaders before independence. Indian leadership had gone along with the principles of nonviolence not out of conviction, but out of convenience. Indian leaders realized that nonviolence was the most effective way of attaining independence without losing life and property. But once independence was achieved, these same leaders wanted nothing to do with nonviolence. Indian leadership was intent on replacing British Imperialists with Indian Imperialists. Unfortunately, we now still have a dominated society – one that is dominated by the political and economic class who have become the modern Maharajas ruling over their fiefdom.

Had Gandhi lived, he would have insisted that bureaucrats and politicians live simply and convert palaces into public facilities such as hospitals and schools. He would have rid the nation of the pomp and pageantry attached to official life. He would have fought against corruption and brought morality and ethics back into public life. Gandhi would have ensured that India had true democracy that was built upward from the grassroots instead of from astroturf. All of this was very inconvenient for those who aspired to be powerful and dominant.

Ultimately, I believe sustainability is very directly linked with world peace. The reason we have not been able to achieve world peace is that we do not know what we want, and so we have attempted half-hearted, unsustainable approaches. World peace must be built painstakingly by dismantling all the nefarious institutions that the culture of violence has spawned. Only by replacing them with a more positive and constructive culture of nonviolence can sustainability become possible – and it will become synonymous with world peace.

***

Arun Gandhi *is the fifth grandson of India's legendary leader, Mohandas "Mahatma" Gandhi. He is founding director of the Gandhi Worldwide Education Institute. He is the author, co-author, or editor of several books, including* A Patch of White *(a book about life in prejudiced South Africa);* Legacy of Love: My Education in the Path of Nonviolence; World Without Violence: Can Gandhi's Vision Become Reality?, *and* The Forgotten Woman: The Untold Story of Kastur, the Wife of Mahatma Gandhi.

Chapter 10
Recycling Reinvented: Music and Sustainability

José-Luis Novo

Igor Fyodorovich Stravinsky, one of the most influential 20th-century composers in the post-Romantic Russian nationalistic era, famously said, "I know that the 12 notes in each octave and the variety of rhythm offer me opportunities that all of human genius will never exhaust." Stravinsky understood and appreciated the challenge of constraints – which may explain why his music went through major shifts in style from the early Russian nationalistic phase, through the neoclassical period, to the ventures exploring serialism.

> Recycling of musical material is one of the best examples of practicing sustainability. It involves creating something new from the old with a healthy dose of imagination, novelty, and discipline.

If you look at Bach, Beethoven, Mozart, and Haydn, it's clear that, like Stravinsky, they knew when to follow the cultural conventions – and when to play around them. This enabled these musical giants to produce fresh musical insights and to make powerful statements that have sustained their music for centuries. Ultimately, I believe it was their resourcefulness to work under constraints and their discipline to produce new works that enabled them to go through so many changes in their musical lives. The message here is that *sustainable development requires resourcefulness and discipline*.

For sustainable development to occur, it should be natural and intuitive. For example, if you examine Bach's music, you will discover that many of his works are loaded with scientific information – I'm not sure if he was aware of it while writing his music. Perhaps it tells us more about his active, experimental mind. Or maybe, a scientific way of thinking was so natural to him and embedded within him that it naturally came out in his scores.

That's the level of awareness and intuitiveness we need to reach to appreciate and practice sustainable development.

In music, as with any work of art, the basic material is transferred from one generation to another. Artists work with what they have inherited – they then enhance their inheritance with creativity and novelty. From a creative perspective, skillfully combining old with new is very difficult and requires imagination and discipline. What's discarded by the previous generation could be up for grabs by the next generation for "artistic recycling."

Consider a recent concert, *Recycling Redefined*, that I conducted for the Annapolis Symphony Orchestra. We performed a masterful, 18-min piece of Brahms' *Variations on a Theme by Haydn* (Op. 56a) and Rachmaninoff's *Symphonic Dances* (Op. 45). After taking a closer look at these scores, it becomes clear that Brahms and Rachmaninoff *recycled* some of their themes from either tradition or previous compositions – yet the way they incorporated previously existing materials with their newly created works is so organically integrated that for an amateur ear it is impossible to differentiate which materials are old and which ones are new. The simultaneous levels of complexity and simplicity in these pieces illustrate how the composers were able to seamlessly swing between the precise and the vague, at the same time offering new insights into an existing work of art. Great artists like Brahms and Rachmaninoff recycled not only the *resources* from previous scores but also certain *features* that people liked about them. That's what makes music sustainable.

Music has quite a bit of connection with tradition. There's so much in music that doesn't belong originally to music. That is, music in large part reflects the composers' thoughts, ideas, emotions, world views, and state of the mind. Music is also an extraordinary language that helps connect with both abstract thought and emotion – that's why it touches so many people at so many levels.

> Music is the sustainable transfer of emotions from one person to another and allows for the migration of creative emotional ideas from one generation to another.

Music can also help serve as a powerful language for peace and diplomacy. Words, especially hurtful ones, remain permanent, whereas music and its associated subjectivity weave together, producing an enduring experience. Consider, for example, the New York Philharmonic Orchestra's diplomatic trip to North Korea under the leadership of Maestro Lorin Maazel in 2008. Music can create a synergy that's much more harmonious than a speech, which, of course, can be interpreted in so many different or extreme ways. German composer Felix Mendelssohn from the early Romantic era captured this view best: "Even if, in one or other of them, I had a particular word or words in mind, I would not tell anyone, because the same word means different things to different people. Only the songs say the same thing, arouse the same feeling, in everyone – a feeling that can't be expressed in words."

Any composer tries to add a different type of dimension and freshness to existing musical material. During that process, the composer creates new sound waves, emotions, and insights that affect the current generation of people and the ones to come. Music, much like waves, travels from one place to another, one culture to another, and one generation to another. In ancient cultures, a good memory helped sustain the oral tradition of education and wisdom. Our society is now hyperactive, with decreasing attention spans. And the music of our generation – which is generally noisy and remixed – is reflective of our busy minds. The brain may adapt to fast-moving circumstances – but to what extent and at what cost? It is certainly not a healthy way. The best thing we could do for our children is to plant the seeds of artistic creativity and awareness in them early on. That's one way to think about and influence sustainable development.

> Music can be a powerful diplomatic tool. It can create a synergy that's much more harmonious than a speech.

Sustainability, as a concept, idea, or ideal, has different components and interpretations. It doesn't matter whether sustainability is embedded within a culture or it is instead imposed upon that culture. Ultimately, creating something new out of existing material is the basic concept of music and evolution. That's also how I would define progress in sustainable development. You have to find a way of doing something new with what you have. And music has a lot to offer in that regard.

***

Maestro José-Luis Novo *is the James W. Cheevers Music Director Chair of the Annapolis Symphony Orchestra in Maryland. He is also the music director and conductor for the Binghamton Philharmonic in New York. He began his musical studies at the conservatory of Valladolid – his hometown in Spain – obtaining the degree of Profesor Superior de Violín with honors in solfege, harmony, and violin. Novo continued his studies at the Royal Conservatory of Music in Brussels, where he earned a first prize in violin. He was a Fulbright Scholar, obtaining both Master of Music and Master of Musical Arts degrees from Yale University, where he was also bestowed the Frances G. Wickes Award and the Yale School of Music Alumni Association Prize. He completed a Master of Music degree in orchestral conducting at the Cleveland Institute of Music, and concluded his conducting studies at the University of Cincinnati College-Conservatory of Music. Novo has received a number of artistic achievement awards and has led more than a dozen world premieres of commissioned compositions.*

Chapter 11

Connectivity and Sustainability: Perspectives from Landscape and Urban Design

Diana Balmori

If you ask people about what we who work on urban landscape and urban design do, most will say that we deal with planting. And we do. But that vision belongs to the 1900s, when landscapers were reduced to shrubbing up a site around an architect's building. However, since the 1980s, because of the growing interest in the health of the planet, landscape has dealt with soil, air, water, microclimate, wildlife, and (unexpectedly) cities, places of concentrated human habitation. So landscape has moved far beyond plants. Most important of all, we now understand that all the elements of a landscape are interdependent; without dealing with the interconnectedness of all its parts, it is not possible to maintain the health of any of them. But the interdependence goes even further. Landscape's scope reaches beyond the insects, birds, and other creatures that plants attract, and beyond the microclimatic effect of a particular planting. Landscapes can also have a cumulative, mitigating effect on the pattern of heat islands – where temperatures are dramatically higher than in areas of open land – of entire urban areas. The microclimate of a landscaped building is connected to that of the whole city.

This connectivity is central to sustainability.

Consider this example: A vegetable skin, or living roof, is a series of layered filters above a shallow reservoir of water and under a shallow cover of light soil for the planting. In this modest landscape intervention, the plants cool the area they cover and store and clean rainwater. They also keep storm water from entering streams during heavy downpours, when those streams would otherwise overwhelm the drainage

> Sustainability means connectivity and beauty. Only a space that is beautiful can deeply affect and engage its users. Thus, only a space that is truly beautiful is truly lasting.

Fig. 11.1 Linear parks give rise through their continuity to a much greater richness of species (Photo courtesy of Farmington Canal Greenway Master Plan)

system and cause sewage to overflow into rivers. Thus, even a modest vegetation cover on an urban roof is connected to the health of rivers.

There are many more examples of the importance of landscape connectivity. Linear corridors become refuges for plants and animals in isolated open fields. The continuity of these corridors gives rise to a much greater richness of species than that found in the surrounding areas (see Fig. 11.1).

Planting urban trees together in a continuous trench (rather than alone in individual plots) also has beneficial effects. As the U.S. Department of Agriculture's tests in the 1980s revealed, the intermingling of roots in the trench contributes to the higher survival rate and better health of the trees (see, for example, Fig. 11.2).

The fracturing of formerly continuous landscapes has led to dwindling populations of various species and other forms of life connected to them. Now, however, there is new interest in creating bridges for wildlife across highways, to counteract that process. This is not to say that connectivity will always be an asset; we must beware the spread of disease among plants – a case when isolation is beneficial. Nonetheless, overall there is a compelling need to create new connections – such as those made by green roofs and the other landscape interventions discussed above – that contribute to the vitality and health of the environment as a whole.

But the basis for stressing this connectivity is that we now view nature as an ecosystem – that is, as a web whose essence lies in the interdependent connections among the physical and biological systems of our planet. The term "ecosystem" was coined in 1866 by Ernest Haeckel, a German

Fig. 11.2 Continuous trenches for planting urban trees contribute to their higher survival rate and better health. *Left line*: trees planted individually. *Right line*: trees planted in a continuous trench (Photo courtesy of Urban Horticultural Institute, Cornell University)

> "Sustainability" seems to be a vague, bureaucratic term that has to be explained over and over again. "Connectivity" is very specific and captures what is at the heart of the relationship among all things within our new view of nature as ecosystem – it is the essence of urban landscape design.

marine biologist, but it was not understood until the 1980s – more than a century later – when it was embraced by many different disciplines as an accurate model of our world. Moreover, today we view ourselves as part of nature. The work of Darwin, Haeckel, and other biologists has revealed, little by little but with great clarity, that we are all part our planet's ecosystem. For people in the 1700s, however, the world consisted of two separate realms: the realm of nature – which included the physical environment and all the creatures living in it – and the realm of human beings. People at that time also believed that they could shape and use nature without affecting the human realm. Dividing the world in that way therefore allowed them to treat it without regard to the consequences of their actions. During the Industrial Revolution, for example, we favored the human sphere at the expense of other elements of the world's ecosystem. Today, however, our interests and attitudes are adjusting our understanding of ourselves as dependent upon and woven into nature's web.

Comprehending this connectivity has thus changed our relationship with the rest of the world. We now appreciate the importance of sustainability – that is, of keeping the whole ecosystem alive, healthy, and able to continually renew itself. For urban landscape work, that means looking for ways to support the vitality of the ecosystem and helping to create critical connections among all its parts.

It may come as a surprise that this connectivity at the heart of sustainability transforms those who work with connectivity and sustainability into activists. This activism takes many forms. In my own case, it has meant writing *A Landscape Manifesto*. Perhaps many others who are concerned with sustainability are similarly motivated. Now that we know more about the structure of nature, we have to move beyond our old understandings and act as new kin to the rest of the world.

> We need to give visual examples of sustainability that people can relate to and connect with.

***

Diana Balmori *is founding principal of Balmori Associates, a New York City–based landscape urban design firm. Balmori was appointed a Senior Fellow in Garden and Landscape Studies at Dumbarton Oaks in Washington, DC. Balmori is currently serving her second term on the U.S. Commission of Fine Arts. A design educator as well as practitioner, Balmori teaches at the Yale School of Architecture, where she was recently the William Henry Bishop visiting professor of architectural design.*

Chapter 12
Nutrition and Sustainability

Marc Van Ameringen

I am the Executive Director of the Global Alliance for Improved Nutrition (GAIN) and have worked in the field of development and human nutrition for a large part of my professional life.

More than two decades ago, at an event almost forgotten – the 1990 United Nations' Children's Summit – a renewed effort began to address the problem of vitamin and mineral deficiencies, often referred to as "hidden hunger," which today *still* blights the lives of some two billion people. As a result, programs to deliver micronutrients critical to health – especially Vitamin A, iron, iodine, zinc, and folic acid – and which many adults and children lack in their diet – were launched using health and aid systems to get these nutrients to millions of the world's poorest and most vulnerable.

The two decades since have radically changed thinking about both the scale of the undernutrition challenge and its importance to every aspect of human development. I will say more about this later.

> Sustainability, in the context of the food, is about reengineering market production and consumption so that everyone has regular access to affordable, nutritious foods.

But, as importantly, we have also learned a lot about how to sustainably and cost-effectively eliminate the causes of this crisis.

The core message is that adequate nutrition – like all the fundamentals of a safe and healthy life – can't be an "add-on," a compensation or remedy for a poor basic diet. It needs to be built into the very DNA of the way we produce, distribute, and consume food, delivering a proper diet for everyone, not just a few. To do this requires innovation in the way food is produced and consumed. This may be a simple point, but it leads to a radical shift in thinking about sustainability, food security, and health.

Let me illustrate this through a simple scenario. If a child or a poor person lacks the essential vitamins and minerals to combat disease, grow normally, and prosper, we can in many cases fix this through the health and welfare system, buy the necessary supplements, deliver them to clinics, and distribute them through social programs. This certainly can work, but it can be costly and difficult, and it only works so long as a government health department or a foreign donor can finance it. Too often, aid moves on, or an economic crisis squeezes the health budget.

But what if many of these vital vitamins and minerals could be included in everyday products available at the local shop or corner store, were affordable and accessible, and were part of an everyday diet? Governments would not need to finance it, but rather the shopkeeper would just stock the product, and the normal commercial market processes would ensure consumers enjoyed an improved diet.

This scenario is, of course, an actual example: One of India's top three food manufacturers, Britannia Industries, which began manufacturing iron-fortified biscuits for World Food Programme school feeding programs, recently expanded fortification across all its commercial baked product lines, adding iron, iodine, zinc, and Vitamins A and B to, and removing *trans* fats from, its dairy products. Britannia sells 3.6 billion packets of biscuits per year. These biscuits are much more acceptable to consumers than oral iron supplements, with almost a 100% uptake. By building this fortification into the basic operating model, the regular market does the heavy lifting, not the government or doctors and nurses, who can concentrate on other health challenges. The product is integrated into Britannia's business plans, part of the "bottom line," and thereby robust and sustainable.

There have been several critical drivers of this success story. First, Britannia had in GAIN a partner able to provide technical input and help it map the likely (expanding) size of the market for healthier products. Second, the company was able to benefit from a positive environment around health messaging – itself reinforcing messages about the importance of avoiding iron deficiency, creating an opening for a private company to include this in its marketing. This positive demand environment is critical because it changes the consumer demand profile for products.

> Sustainability demands that we work out not just what works for now, or because of exceptional short-term efforts, but is a systematic response that will work for future generations. Donors in particular need to heed this.

But, most importantly, Britannia identified the commercial opportunity to advance its brand identity and loyalty and thereby expand sales for its products. This incentivized the company to find ways to absorb the relatively small cost of fortification, improving its efficiency and cost control. Expanding and shaping this demand and achieving commercial success became a further driver for the strategy and for the health approach to be integrated across its products.

Britannia is not alone; many companies in the food and beverage sector, whether national, regional, or multinational firms, are experimenting with different approaches to improve the nutritional content of their products and in developing new business models to service the base of the economic pyramid. Most of these companies are part of the GAIN Business Alliance, which provides a noncompetitive platform for companies to learn from each other as they begin delivering new offerings. There are many areas of learning so far; one is that building consumer awareness is critical to the success of this new approach, and this in turn creates a fertile area for companies to work with governments, nongovernmental organizations, and civil society to promote healthier diets. There is a clear need for more business-to-business collaboration and public–private partnerships to expand these approaches to a scale where they can have a significant impact on the burden of malnutrition in the world today.

It is important not to overemphasize this new approach to addressing malnutrition. The delivery of nutritional interventions through public and social systems remains fundamentally important to achieving impact; however, it is equally important to begin to see markets as a critical part of the solution. Food is quite different from public goods such as medical care and education – where government normally funds and leads service delivery.

Almost all food is produced and distributed by the private sector – farmers, traders, and food companies. Making food economies more "nutrition-friendly" requires innovation to link health and public services, the producers, and the consumer. Implementation requires understanding the impact of market forces on the poorest and finding solutions to shape the market to benefit improved nutrition: through new and improved food products, better targeting of those in need outside the market, and innovative messaging to consumers.

GAIN was set up to develop this thesis in practice. In the past 10 years, GAIN has developed many market-driven initiatives, which bring government and producers together with civil society to regulate, produce, and market beneficial fortified staples. Today, GAIN delivers nutritionally enhanced products to an estimated 610 million people in more than 30 countries. These products are as

> It is important to understand that durable, long-term solutions usually require governments, business, and civil society to work together.

varied as fortified cooking oil, flour, soy sauce, and even biscuits. GAIN also intervenes to protect the most vulnerable segments in the society, namely, women, infants, and young children or those affected by emergencies or chronic illnesses. Our practical experience has been obtained through programs with hundreds of government bodies, community groups, researchers, and the private sector in every continent. Through this work, GAIN aims to make good nutrition a permanent characteristic of the global food economy, by calling together all essential stakeholders in the fight against malnutrition.

The challenges are many. For instance, GAIN's Maternal, Infant, and Young Child Nutrition (MIYCN) programs approach clearly illustrates the sustainability of the products and interventions selected. Because of their lifelong impact, there is recent agreement that interventions targeting the nutritional status of women of reproductive age, pregnant and lactating women, and children under 2 need to be prioritized. Such interventions aim to prevent and treat malnutrition, including micronutrient deficiencies, and enable better care for women and children, including exclusive breastfeeding for the first 6 months of life. The supply of nutritious foods is not currently accessible for many households. Hence, food-based interventions for young children (from 6 months onward) require new and more highly nutritious commodities that can complement their daily diet, such as micronutrient powders, lipid-based nutritional supplements, and high-energy cereals/foods. Few affordable products – perhaps less than 2% of those required, are available to those in need, either to be purchased by the consumer or delivered via aid channels.

How will this address undernutrition, hunger, and make our world more stable and fair? Let me recap some of the main changes in the global thinking on nutrition.

First, 2010 was in many ways a decisive year for human health, crystallizing for the first time a global consensus that poor nutrition is the Achilles heel of human development. The 2010 G8 meeting in Canada and the UN Millennium Development Goal (MDG) Summit in September 2010 identified that underinvestment in nutrition is impeding virtually every MDG, from eradicating extreme poverty and hunger, to child survival, addressing underweight and stunting, to cognitive development, to being able to stave off or overcome infectious diseases, to ensuing maternal and newborn child health.

Second, medical science now confirms the period from prepregnancy and conception to the first 2 years of life is a finite "one-chance" window for cost-effective nutrition interventions that reduce mortality and avert permanent damage in growth and cognitive development. This "can't go back" or "address later on" reality means we either make a difference in this window of opportunity or risk impairing the individual's potential, along with the potential (social and economic) of communities and entire countries. Poor early nutrition leaves the individual open to a host of diseases, turning common ailments such as colds and diarrhea into killers. Significantly too, early setbacks are irreversible and can cut an individual's lifetime earnings by up to 10%. Poor nutrition and dietary deficiencies in iron, iodine, and zinc can lower GDP by 2–3%. The conclusion is clear: Unless you improve nutrition, the Millennium Development Goals (MDGS) are beyond reach.

What are the implications of this for sustainability?

First, in food systems, public–private partnerships are critical to deliver sustainable interventions. GAIN mobilizes different stakeholders – governments, nongovernmental and international organizations, businesses, and civil society. Effective delivery requires an approach that understands the potential of linking public sector development and regulation of food and health policies to food production and consumption. These can be knitted together in national strategies that give consumers better information, more choices, and improved food products, harnessing the creative energy of business to make markets more "nutrition-friendly." Producing more nutritious and culturally accepted food products is an important priority and can help target those in need. In China this can mean fortified soy sauce, in West Africa fortified cooking oil, in South Africa flour. Developing Sprinkles or micronutrient powders that can be added to food at the household level is an exciting new approach to target those most in need.

Second, that successful and tangible development intervention requires the leadership of the public sector and government. The right delivery of improved nutrition to the most vulnerable needs to be instituted in a coordinated national plan since successful interventions in nutrition are multisectoral in nature, that is, linking together basic food and nutrition security components, while sharply improving the nutritional aspects of agriculture, health, and education. GAIN works closely with national governments to ensure the political buy-in and strengthen the public sector's role in improving nutrition – an absolutely fundamental step to implementing comprehensive and efficient nutrition interventions.

Third, innovation is critical to sustainability. GAIN is founded on a belief that the private sector and market innovation within a publicly managed health and food security system can unleash a multitude of new ways to improve nutrition security. The development of new products and marketing approaches takes time and effort. A structured partnership with business helps fast-track new business investments. This enables exchange between public agencies and large and small businesses – to discuss and pilot new business models and improve dialogue between policy makers and businesspeople. GAIN's Business Alliance is the global focal point for private sector partnerships around undernutrition in Africa, South Asia, Southeast Asia, and Latin America and the Caribbean, mobilizing business across the nutrition value chain.

Finally, we have the importance of advocacy: Sustainability is not a solution in itself. This is an important reflection for all: Sustainability is a concept that is generally aligned with calls for radical changes in the way we live and act, whether it be in relation to energy, climate change, food, or resource management. But a solution is only relevant when it is recognized that the problem has to be addressed! Advocacy is an important part of creating demand for change – campaigning, education, and priority for nutrition to establish the paradigm of a sounder global nutrition regime. Through its national and global advocacy work, GAIN fosters improved policies and practices among governments, international organizations, and national and global nutrition landscapes. Sustainability in the case of a basic right such as food security and good nutrition therefore has a clear normative or ethical dimension: It needs to harness willpower within all sectors of society to give priority to finding practical solutions.

Reflecting on sustainability is as old as human society, even if it is a modern concept. It is about using experience to define a problem, and finding the way to use collective willpower, leadership, and resources to fix it, not just for now, but for the long term. As a campaign on nutrition, GAIN seeks solutions that are workable and permanent and that contribute to this betterment of humanity through diet. If the improvement of the human condition in the 19th century was marked by the new understanding of environmental health – of sewers and clean water – and the 20th century by antibiotics and eradication of smallpox, the 21st century can be about creating a human-friendly food system. That is GAIN's goal, and the vision of sustainability in the food sector we promote.

***

Marc Van Ameringen *is the Executive Director of the Global Alliance for Improved Nutrition (GAIN), which supports programs aimed at reducing malnutrition with a focus on micronutrient deficiencies. Prior to joining GAIN, he was Vice President of the Canada-based Micronutrient Initiative, where he was responsible for coordination. Earlier, he was Special Advisor to the G8 Summit within the Canadian Government's Department of Foreign Affairs and International Trade, and a director based in Africa for the International Development Research Centre (IDRC). He serves on a number of boards and advisory bodies involving public and private sectors in health and nutrition.*

Chapter 13

A Poet in the Car Company: Sustainability of Passion and Profitability

David Berdish

When Bill Ford was named chairman of the board for the Ford Motor Company in 2000, I was honored to be selected as Ford's global manager of Human Rights. I spent my first 17 years at the company working in manufacturing – often in jobs that I hated but loving the people I worked with. I often joke that I'm probably the only person in Ford that has a bachelor's degree in poetry and a graduate degree in operations research, but I think this unusual background put me on a team that required a different type of systems thinking.

The toughest part of my job wasn't dealing with complex production systems – it was working on processes that enable teams of talented people to accomplish much more than what they could do by working alone. My work in human rights helped demonstrate that an organization isn't only an entity of economics, strategy, and technology – it is also a community of people.

I had the coolest job in the world. But just as I began to rock and roll in my new responsibility, my life came to a crashing halt. After a six-year battle with cancer, my wife, Renee, died and left my 8-year-old son and me alone, utterly confused and heartbroken. Drew and I had to deal with the pain, but I also needed to make sure that there were love and normalcy in his life, which meant carpools, nannies, sports, and home management while I tried to lead an important sustainability initiative in a demanding corporate environment.

> Sustainability should not be thought of as an entity of economics, strategy, and technology – it should be about people.

Renee's death forced me to reorient my perspective about the way I lived and worked. I embraced my role as a father in an entirely different way; I let my love and faith drive my decision making. So, I honestly approached my to-do list and removed those tasks that I developed to increase my visibility and recognition. I avoided petty squabbles at the office. I even had the courage to decline meetings that I felt were unnecessary to accomplish my sustainability objectives and tried to stay out of the fray as fellow managers behaved like every meeting was a life-and-death situation. I saw life and death, and what happens at Ford meetings doesn't even come close.

My changing perspective and my deeper sense of love for my son kept me on focus. I even think that being a widower forced me to add some "Mom" roles, which (I hope) helped to nurture Drew and think more often about the consequences of my behavior and actions around him.

I am convinced that my new sense of balance and priority helped me lead Ford in several ambitious human rights objectives, including the implementation of the Ford Code of Basic Working Conditions (the first of its kind in the industry); the first automaker to release details of how the HIV/AIDS epidemic affects the corporation; a policy letter that sets forth the company's guiding principles for labor and environmental standards throughout its global operations; and the only automaker and heavy manufacturer to participate in the United Nations Global Compact Human Rights Working Group.

I worked hard to be trustworthy and developed an incredibly close relationship with the Interfaith Center on Corporate Responsibility. ICCR helped me with decisions and challenging issues around water, conflict minerals, and trafficking. How can one not become passionate about that with so much at stake?

> Improved sustainable performance is not just a requirement, but a tremendous business opportunity.

Of course, being passionate and showing courage in decisions is not enough, even with a deep understanding of the subject. A great company achieves sustainability goals through leadership. In 2007, when Ford Motor Company was struggling with its financial performance and handing out severance packages, I was nervous in how my work would be judged when the Ford culture was in survival mode.

I didn't need to worry, as I was blessed with a wonderful director, John Viera, and group vice president, Sue Cischke. Viera and Cischke brought credibility to our work that was until then often perceived to be little value-add staff work.

Viera was chief engineer of the Navigator-Expedition and knew inner workings of product development. He launched one of the hated icons of environmental leaders, so he knew what he was into but engaged in a thoughtful and empathetic way, both internal and external to the company. Prior to 2007, my bosses had passion and the best of intensions, but their experiences were in staff functions. Because Viera was successful at "running the business," he was not ignored as a "staffie." And NGOs were more confident that Viera could represent our company when it came to fuel economy, design, and product strategy.

Cischke is an engineer, too, and was already in a challenging executive position that is responsible for the environmental and safety standards of our vehicles. She made the connection to sustainability, took the strategy seriously, and directed the objectives with deep knowledge, understanding, and respect.

Further, Viera and Cischke were empathetic of my personal challenges. Each expected results, but I was given autonomy to set directions whether I was at World Headquarters, a college campus, or my home office. My creativity was fully charged, and with that type of support, I felt that we could accomplish anything together in climate change, safety, and reporting – and that I was empowered to work on the social dimensions of sustainability – which at Ford meant a focus on human rights, water, and sustainable mobility.

> It's impossible to design a sustainable solution in the usual linear analytic way.

For example, Ford continues to reduce carbon emissions and is pressed to produce electric vehicles; the added cost of this technology is not only the price differential of a battery.

The impact of lithium – the dangerous working conditions, the impact on indigenous populations, and the water used to extract the raw materials – are all system variables that need to be balanced.

As we design our products and develop markets with respect to electrification and information technology, there will be impacts on our business practices, supply chain, and policy statements – including a wide variation in working conditions and labor laws in emerging markets – especially with respect to "conflict minerals," women, trafficking, child and slave labor. So the work is important to secure license to operate in low-cost markets, and our trustful and credible relationship with human rights groups ensures quiet resolution of potential Ford-related issues.

Cischke notes that "Water issues are increasingly important to our stakeholders, including our customers, investors and business partners. Water conservation and greenhouse gas reduction are integral to Ford's sustainability strategy. By reporting on them, we support positive social change and encourage the reduction of the environmental impact of our facilities." So we are studying the analysis of Ford "splash," including life-cycle analysis, use in direct operations, and use in value chain. We are working with global stakeholders to identify water-scarce regions and areas of concern and to support community water efforts.

When Bill Ford speaks, he often reminds us that improved sustainable performance is not just a requirement, but a tremendous business opportunity. Ford, the U.S. Department of State (State), and George Washington University (GW) have formed a partnership with the objective to leverage the mobility of Ford's SYNC technology and the connectivity of the IT "cloud," the diplomatic and policy relationships of State, and the public–private partnership expertise of GW's Institute of Corporate Responsibility to launch a pilot project providing access to basic healthcare services to women in underserved areas in and around Chennai, India. The partnership seeks to explore a new framework for successful public–private partnerships based on trust.

I have been able to develop my skills as a result of personal development, my perspective, and, most of all, my relationships. My friends and colleagues have taught me to better understand diverse cultures and the global challenges of sustainable development. More importantly, my life experiences have taught me that the challenges in life and sustainability can only be addressed if we stop looking at sustainability as a series of financial decisions, but a complex web of social, environmental, and economic systems in which global companies need to operate in order to create long-term

> *If I could do it again, then*
> - More operational people would be involved. Development of strategy, policy papers and communication are important but could have been done more quickly and with fewer guru and staff types.
> - I would learn about our employees' passion. I wish I could have taken inventory in all of the organizations in the company, as people were doing incredible things as part of their jobs or on their own time to contribute to a sustainable Ford. I would have used more of those stories as the foundation of our sustainability principles, and relied less on the gurus.
> - We would have designed application projects earlier. Companies like Ford want to see results. There were several grassroots sustainability efforts going on in the organization that could have been supported, documented, and communicated.

success and profitability. And the human elements of those systems, including working conditions, access to clean water and education, healthcare and other related issues, should be central to businesses' sustainability strategies.

***

David Berdish *is the Manager of Social Sustainability at Ford Motor Company. He has been at Ford since 1983 and has worked in Production, Program Management, Finance, and Organizational Learning. Berdish is responsible for the Ford Human Rights Code of Working Conditions and the sustainable water strategy and is the program manager to understand urban markets, megacity mobility, and nontraditional transport. He has worked with cities in the United States, South Africa, India, and Brazil to integrate information technology, infrastructure, and the role of alternative and electric vehicles into sustainable mobility solutions. Berdish received his B.A. from the University of Michigan and his M.S. from Virginia Commonwealth University.*

Chapter 14
The Need for Sustainable Heretics

Freeman Dyson

In the modern world, science and society often interact in a perverse way. We live in a technological society, and technology causes political problems. The politicians and the public expect science to provide answers to the problems. Scientific experts are paid and encouraged to provide answers. The public does not have much use for a scientist who says, "Sorry, but we don't know." The public prefers to listen to scientists who give confident answers to questions and make confident predictions of what will happen as a result of human activities. So it happens that the experts who talk publicly about politically contentious questions tend to speak more clearly than they think. They make confident predictions about the future and end up believing their own predictions. Their predictions become dogmas that they do not question. The public is led to believe that the fashionable scientific dogmas are true, and it may sometimes happen that they are wrong. That is why heretics who question the dogmas are needed.

As a scientist, I do not have much faith in predictions. Science is organized unpredictability. The best scientists like to arrange things in an experiment to be as unpredictable as possible, and then they do the experiment to see what will happen. You might say that if something is predictable, then it is not science. When I make predictions, I am not speaking as a scientist. I am speaking as a storyteller, and my predictions are science

> The world always needs heretics to challenge the prevailing orthodoxies. Since I am a heretic, I am accustomed to being in the minority. If I could persuade everyone to agree with me, I would not be a heretic. We are lucky that we can be heretics today without any danger of being burned at the stake.

fiction rather than science. The predictions of science fiction writers are notoriously inaccurate. Their purpose is to imagine what might happen rather than to describe what will happen. I will be telling stories that challenge the prevailing dogmas of today. The prevailing dogmas may be right, but they still need to be challenged. I am proud to be a heretic. The world always needs heretics to challenge the prevailing orthodoxies. Since I am a heretic, I am accustomed to being in the minority. If I could persuade everyone to agree with me, I would not be a heretic.

We are lucky that we can be heretics today without any danger of being burned at the stake. But unfortunately I am an old heretic. Old heretics do not cut much ice. When you hear an old heretic talking, you can always say, "Too bad he has lost his marbles," and pass on. What the world needs is *young* heretics. I am hoping that one or two of the people who read this piece may fill that role.

Several years ago, I was at Cornell University celebrating the life of Tommy Gold, a famous astronomer who died at a ripe old age. He was famous as a heretic, promoting unpopular ideas that usually turned out to be right. Long ago I was a guinea pig in Tommy's experiments on human hearing. He had a heretical idea that the human ear discriminates pitch by means of a set of tuned resonators with active electromechanical feedback. He published a paper in 1948 explaining how the ear must work. He described how the vibrations of the inner ear must be converted into electrical signals that feed back into the mechanical motion, reinforcing the vibrations and increasing the sharpness of the resonance. The experts in auditory physiology ignored his work because he did not have a degree in physiology. Many years later, the experts discovered the two kinds of hair cells in the inner ear that actually do the feedback as Tommy had predicted, one kind of hair cell acting as electrical sensors and the other kind acting as mechanical drivers. It took the experts 40 years to admit that he was right. Of course, I knew that he was right, because I had helped him do the experiments.

Later in his life, Tommy Gold promoted another heretical idea, that the oil and natural gas in the ground come up from deep in the mantle of the earth and have nothing to do with biology. Again, the experts are sure that he is wrong, and he did not live long enough to change their minds. In 2004, just a few weeks before he died, some chemists at the

> Climate change is a contentious subject, involving politics and economics as well as science. The science is inextricably mixed up with politics. Everyone agrees that the climate is changing, but there are violently diverging opinions about the causes of change, about the consequences of change, and about possible remedies.

Carnegie Institution in Washington did a beautiful experiment in a diamond anvil cell. They mixed together tiny quantities of three things that we know exist in the mantle of the earth, and observed them at the pressure and temperature appropriate to the mantle about 200 km down. The three things were calcium carbonate, which is sedimentary rock, iron oxide, which is a component of igneous rock, and water. These three things are certainly present when a slab of subducted ocean floor descends from a deep ocean trench into the mantle. The experiment showed that they react quickly to produce lots of methane, which is natural gas. Knowing the result of the experiment, we can be sure that big quantities of natural gas exist in the mantle 200 km down. We do not know how much of this natural gas pushes its way up through cracks and channels in the overlying rock to form the shallow reservoirs of natural gas that we are now burning. If the gas moves up rapidly enough, it will arrive intact in the cooler regions where the reservoirs are found. If it moves too slowly through the hot region, the methane may be reconverted to carbonate rock and water. The Carnegie Institution experiment shows that there is at least a possibility that Tommy Gold was right and the natural gas reservoirs are fed from deep below. The chemists sent an e-mail to Tommy Gold to tell him their result and got back a message that he had died 3 days earlier. Now that he is dead, we need more heretics to take his place.

The main subject of this essay is the problem of climate change. This is a contentious subject, involving politics and economics as well as science. The science is inextricably mixed up with politics. Everyone agrees that the climate is changing, but there are violently diverging opinions about the causes of change, about the consequences of change, and about possible remedies.

Allow me to promote a heretical opinion: All the fuss about global warming is grossly exaggerated. Here I am opposing the holy brotherhood of climate model experts and the crowd of deluded citizens who believe the numbers predicted by the computer models. Of course, they say, I have no degree in meteorology and I am therefore not qualified to speak. But I have studied the climate models and I know what they can do. The models solve the equations of fluid dynamics, and they do a very good job of describing the fluid motions of the atmosphere and the oceans. They do a very poor job of describing the clouds, the dust, the chemistry and the biology of fields and farms and forests. They do not begin to describe the real world that we live in. The real world is muddy and messy and full of things that we do not yet understand. It is much easier for a scientist to sit in an air-conditioned building and run computer models than to put on winter clothes and measure what is really happening outside in the swamps and the clouds. That is why the climate model experts end up believing their own models.

There is no doubt that parts of the world are getting warmer, but the warming is not global. I am not saying that the warming does not cause problems. Obviously, it does. Obviously, we should be trying to understand it better. I am saying that the problems are grossly exaggerated. They take away money and attention from other problems that are more urgent and more important, such as poverty and infectious disease and public education and public health, and the preservation of living creatures on land and in the oceans, not to mention easy problems such as the timely construction of adequate dikes around the city of New Orleans.

> When we are trying to take care of a planet, just as when we are taking care of a human patient, diseases must be diagnosed before they can be cured. We need to observe and measure what is going on in the biosphere rather than relying on computer models.

I will discuss the global warming problem in detail because it is interesting, even though its importance is exaggerated. One of the main causes of warming is the increase of carbon dioxide in the atmosphere resulting from our burning of fossil fuels such as oil and coal and natural gas. To understand the movement of carbon through the atmosphere and biosphere, we need to measure a lot of numbers. I do not want to confuse you with a lot of numbers, so I will ask you to remember just one number. The number that I ask you to remember is one hundredth of an inch per year. Now I will explain what this number means. Consider the half of the land area of the earth that is not desert or ice cap or city or road or parking-lot. This is the half of the land that is covered with soil and supports vegetation of one kind or another. Every year, it absorbs and converts into biomass a certain fraction of the carbon dioxide that we emit into the atmosphere. Biomass means living creatures, plants and microbes and animals, and the organic materials that are left behind when the creatures die and decay. We don't know how big a fraction of our emissions is absorbed by the land, since we have not measured the increase or decrease of the biomass. The number that I ask you to remember is the increase in thickness, averaged over half of the land area of the planet, of the biomass that would result if all the carbon that we are emitting by burning fossil fuels were absorbed. The average increase in thickness is one hundredth of an inch per year.

The point of this calculation is the very favorable rate of exchange between carbon in the atmosphere and carbon in the soil. To stop the carbon in the atmosphere from increasing, we only need to grow the biomass in the soil by a hundredth of an inch per year. Good topsoil contains about 10% biomass, so a hundredth of an inch of biomass growth means about a tenth of an inch of topsoil. Changes in farming practices such as no-till farming, avoiding the use of the plow, cause biomass to grow at least as fast as this. If we plant crops without plowing the soil, more of the biomass goes into roots, which stay in the soil, and less returns to the atmosphere. If we use genetic engineering to put more biomass into roots, we can probably achieve much more rapid growth of topsoil. I conclude from this calculation that the problem of carbon dioxide in the atmosphere is a problem of land management, not a problem of meteorology. No computer model of atmosphere and ocean can hope to predict the way we shall manage our land.

Here is another heretical thought. Instead of calculating worldwide averages of biomass growth, we may prefer to look at the problem locally. Consider a possible future, with China continuing to develop an industrial economy based largely on the burning of coal, and the U.S. deciding to absorb the resulting carbon dioxide by increasing the biomass in our topsoil. The quantity of biomass that can be accumulated in living plants and trees is limited, but there is no limit to the quantity that can be stored in topsoil. To grow topsoil on a massive scale may or may not be practical, depending on the economics of farming and forestry. It is at least a possibility to be seriously considered that China could become rich by burning coal while the U.S. could become environmentally virtuous by accumulating topsoil, with transport of carbon from mine in China to soil in America provided free of charge by the atmosphere, and the inventory of carbon in the atmosphere remaining constant. We should take such possibilities into account when we listen to predictions about climate change and fossil fuels. If biotechnology takes over the planet in the next 50 years, as computer technology has taken it over in the last 50 years, the rules of the climate game will be radically changed.

When I listen to the public debates about climate change, I am impressed by the enormous gaps in our knowledge, the sparseness of our observations, and the superficiality of our theories. Many of the basic processes of planetary ecology are poorly understood. They must be better understood before we can reach an accurate diagnosis of the present condition of our planet. When we are trying to take care of a planet, just as when we are taking care of a human patient, diseases must be diagnosed before they can be cured. We need to observe and measure what is going on in the biosphere rather than relying on computer models.

***

Freeman Dyson *is professor emeritus of physics at the Institute for Advanced Study in Princeton, New Jersey. His research has involved work in quantum electrodynamics, nuclear reactors, solid-state physics, ferromagnetism, astrophysics, and biology, looking for problems where elegant mathematics could be usefully applied. Dyson is a fellow of the Royal Society of London and a member of the National Academy of Sciences. Among other honors, he has received the Templeton Prize for progress in religion. He has written a number of books about science for the general public.*

Acknowledgment This essay has been adapted from *Many Colored Glass: Reflections on the Place of Life in the Universe* by Freeman Dyson, University of Virginia Press, 2007.

Chapter 15
Performance with Purpose

Dave Haft

Environmental sustainability has been a top priority for Frito-Lay for nearly 20 years – although, back then, we didn't call it "environmental sustainability" or even realize what an important role we could play in helping our planet.

When we began our environmental efforts in 1993, we called it "compliance" and focused on adhering to federal, state, and local environmental regulations. Recognizing the importance of ensuring every part of our business – which included 40 plants, 200 distribution centers, and 15,000 trucks – met or exceeded environmental regulations, we decided employee ownership was key. So we formed "Green Teams" – groups of employees, including both plant managers and frontline operators, across our facilities, who we trained in areas such as waste water management, air emissions reduction, hazardous material management, and spill prevention.

> Sustainability is about doing the right thing for our planet, for our employees and consumers, and for our company. By protecting the earth's natural resources through innovation and more efficient use of land, energy, water, and packaging, we're helping build a better future for our planet and for our business.

As they learned to measure, analyze, and improve their sites' environmental performance, these employees became highly engaged in and enthusiastic about our environmental programs, and their contributions helped their sites earn top scores in external audits year after year.

Then, in 1997, Frito-Lay's senior Operations leaders launched a visionary productivity program aimed at increasing our efficiency and competitiveness, while reducing our operating costs. The goal of the program, called "Starfleet" because of its vast reach and forward-thinking ideas, was to optimally integrate our people, processes, and technology across our business to deliver as much as $100 million in annual savings.

One of the first Starfleet initiatives focused on reducing our water and energy use. As you can imagine, washing, slicing, cooking, seasoning, and packaging millions of pounds of corn and potato products each day requires a huge amount of energy and water. So the opportunities for cost savings were significant. For the first time, we dedicated a team of engineers – who we nicknamed Frito-Lay's "Department of Energy" – to resource conservation. I was a vice president of operations at the time and volunteered to be the team's executive sponsor.

Inspired by business consultants and authors Jim Collins and Jerry Porras' concept of Big Hairy Audacious Goals, or "BHAGs" – the idea that companies should set visionary, emotionally compelling goals – I proposed reducing our water use by 50%, our natural gas use by 30% and our electricity use by 25%, per unit of production, over the next 10 years. If successful, this would save us $50 million in annual operating costs. Few others at Frito-Lay thought this was even remotely feasible. Yet, compelled by the potential cost savings, senior management gave us the funding to move forward. Indra Nooyi, now the Chairman and CEO of our parent company, PepsiCo, and at the time PepsiCo's chief financial officer, agreed to a slightly lower internal rate of return (IRR) for these projects compared to our standard productivity programs, believing they would pay off in the long term. This kind of confidence and support from our top executives would prove to be critical as we continued our productivity and environmental sustainability journey.

With this endorsement, our "Department of Energy" set to work. We replaced outdated, energy-draining manufacturing equipment with state-of-the-art technology, including high-efficiency lighting, air compressors, and boiler controls. We developed innovative heat recovery systems that captured and reused waste heat from our potato chip and tortilla chip processes. We installed sophisticated co-generation systems, biomass- and biogas-fueled boilers, and solar concentrators and photovoltaic systems. We also installed centrifuges to reduce our reliance on municipal sewage treatment plants, and we began recycling the water we used to process potato chips. The breadth and decentralized nature of our operations gave us an advantage, allowing us to test new technologies and processes at one manufacturing site and, if successful, replicate them at up to 30 other plants.

> Sustainability is about optimizing our processes, technology, and behavior. I'm drawn by the challenge of continually identifying and implementing state-of-the-art systems, processes, and technologies that will allow us to reduce our energy, water, and land use with the goal of having less impact on the earth. I'm encouraged by the idea that educating and empowering our people is ultimately what expands and enhances our environmental sustainability journey.

Sure enough, we were making significant progress toward our "BHAGs." In 2003, the Alliance to Save Energy recognized us as Star of the Year, and in 2006, we received the first of six Federal Environmental Protection Agency Energy Star Partner of the Year awards, including the four most recent of these awards at the prestigious "Sustained Excellence" level.

A few years into our Starfleet program, I took on another challenge after my boss, Frito-Lay's senior vice president of operations, asked me when we would build a "green" facility. Frankly, this was not high on my priority list or even something I'd given much thought to. But after researching the topic and visiting a plant near Detroit that Ford Motor Co. was refurbishing to become a "green" facility, I realized this concept had both environmental and cost-savings potential for Frito-Lay. I promised my boss that the next facility we built would be "green."

In 2005, we cut the ribbon on a state-of the-art "green" distribution center in Rochester, New York. It featured innovations in renewable energy, alternative lighting, recycled building materials, enhanced land management systems, and energy-efficiency standards. The U.S. Green Building Council certified it as a LEED Gold (Leadership in Energy and Environmental Design) site, one of the industry's highest certification levels.

Seeing the financial and environmental benefits, we aligned our senior executives on requiring that all of our new construction be designed for LEED Gold certification. We also secured funding to upgrade our 13 largest manufacturing plants to LEED Gold status. Today, our 700,000-square-foot corporate headquarters building in Plano, Texas, and 13 manufacturing plants throughout the United States are LEED Gold-certified.

Over the next few years, we continued making progress in our electricity, natural gas, and water conservation BHAGs. We got a big boost in 2006, when Al Carey became president and CEO of Frito-Lay North America. When I shared our environmental accomplishments and goals with Al, he told me he firmly believed we were taking all the right steps. He soon made environmental sustainability an integral part of our business priorities.

By 2008, I decided we should add our fleet of delivery and over-the-road trucks to our environmental program. With one of the largest private fleets in the country, the opportunities for fuel and emissions reduction and cost savings were tremendous. I challenged our fleet engineers to their own "BHAG" – a 50% reduction in fuel use by 2020.

Following the approach they'd seen our manufacturing engineers take, our fleet engineers began identifying opportunities for fuel reduction and then developed, tested, and implemented new technologies and improved operating and maintenance practices, considering everything from alternative fuel sources to vehicle design to driver behavior.

They installed idle shutdown systems, auxiliary heaters, and advanced aerodynamic retrofits. They replaced 9-mile-per-gallon (mpg) route delivery trucks with 18-mpg trucks. They also started rolling out all-electric delivery trucks. Knowing that driver habits could significantly impact miles per gallon, we added GPS systems to our vehicles to provide actionable feedback, and we implemented Lean Six Sigma–based driver training programs.

> I would like to see more companies make a big investment in sustainability through the optimal integration of people, process, and technology. As our experience at Frito-Lay and PepsiCo demonstrates, making a serious commitment to reducing our environmental footprint, way beyond regulatory compliance, will enhance bottom-line performance, engage employees, and increase customer and consumer loyalty.

We soon began seeing progress on our fuel BHAG as well, with a 14% reduction in gallons used over 3 years. We've continued our momentum with progress toward a 30-mpg route truck, and we're encouraged by the results of our current propane and compressed natural gas initiatives. We clearly have a line of sight to our goal to reduce fuel use by 50% by 2020.

This additional success gave us even more momentum, and we took on another challenge: becoming a "near-zero-landfill" company. As a packaged-goods company, we'd long known that reducing solid waste was important to our resource conservation goals. For many years, our employees have reused our cardboard shipping cartons five or six times each before recycling them, saving 4.5 million trees and tens of millions of dollars a year. We also improved our snack bags to minimize the amount of packaging film we use, which means we can fit more snack bags on our trucks, thus reducing product delivery miles driven.

In 2008, we decided we needed to do even more to reduce our solid waste and challenged our plants to become "near-zero-landfill" by the end of 2012, sending less than 1% of their manufacturing waste to landfills.

They began recycling food scraps from our manufacturing process to be turned into animal feed, plastics, metals, cardboard, and other materials. Employees across the company took this new challenge to heart and began reducing the number of handouts they distributed at meetings, choosing reusable coffee mugs and utensils over disposable ones, and looking for more opportunities to recycle and reuse materials. Today, 70% of our plants send less than 1% of waste to landfill, and we've reduced our total waste by 80%. We're on track to reach our goal by 2012, and we're expanding the "near-zero-landfill" program to our distribution centers around the country.

We've long since exceeded our original BHAGs, and our resource conservation programs are now saving us over $80 million a year. But I think our greatest opportunities are still ahead. For instance, in 2011, we completed Frito-Lay's most ambitious project yet, retrofitting our 25-year-old Casa Grande, Arizona, plant with state-of-the-art resource conservation technology to take it nearly entirely off the natural gas, electricity, water, and landfill grids. As a result of our work there, this plant now runs on 75% renewable energy, using large-scale photovoltaic single- and dual-axis trackers, Stirling engines, and a biomass boiler. We've also installed a membrane bioreactor with secondary filtration that each day converts up to 75% of 400,000 gal of process water to drinking water that meets all EPA standards. We call this plant our "near-net-zero" site, and we plan to use as much as possible of what we learn here at our other plants. In fact, all of our plants now have a 5-year "near-net-zero" vision that they are executing against. We've also set our sights on becoming the "Pre-eminent Green Company" and have developed a strategy to further our environmental efforts in every part of our business, from seed to shelf – from the agricultural practices of our potato growers to our manufacturing and trucking systems to our packaging materials.

As our environmental efforts have had a positive ripple effect on the business at Frito-Lay – with initiatives spreading to more and more sites and parts of our business, the ripples have also continued to positively impact our people. I hear time and time again from employees of all levels and roles that seeing our environmental commitment at work has inspired them and their families to adapt more sustainability practices at home. And as our sustainability programs proved successful, they have been replicated across many other parts of PepsiCo globally. In fact, Indra Nooyi told Frito-Lay leaders that our commitment to and successes with our environmental programs inspired her to make "environmental sustainability" a key plank of PepsiCo's "Performance with Purpose" vision, which is focused on delivering sustainable financial performance, while also doing the right thing for society, our planet, and our employees.

Of course, just as when we began our environmental journey almost 20 years ago, our bottom line is still a key driving force behind our efforts. But in the course of delivering productivity through resource conservation, we've demonstrated that what's good for our company is also good for the planet and its people.

***

Dave Haft *is Senior Vice President of Productivity, Sustainability & Quality at Frito-Lay, Inc., a division of PepsiCo. During the past 10 years, he has led Operations Productivity, Quality Assurance, Food Safety/Sanitation, Environmental Compliance and Resource Conservation, and Manufacturing Operations Support for Frito-Lay North America. He also has had responsibility for Service and Distribution, including Warehouse, Traffic, Fleet, and Sales Operations.*

Chapter 16

The Business of Sustainability: A Different Design Question

Gregor Barnum

I worked as director of Corporate Consciousness for Seventh Generation – an independent, privately held company based in Burlington, Vermont. The company's products include nontoxic household cleaners, phosphate-free dishwasher and laundry detergents, as well as bleach-free, recycled bath tissue and paper towels. My job was to understand and tackle big challenges as sustainability, systems thinking, and innovation – all engaging topics in themselves. When we were working on a sustainability strategy for Seventh Generation, our objective was simple: to make "less bad" impacts to the environment. This meant reducing our greenhouse gas emissions, reducing our waste impact by adding recycled content to our packaging, even reducing our packaging, and reducing our usage of palm oil, and so forth. Our mantra was *reduce, recycle, reuse.*

This is the framework most companies use. Somehow it has become the standard practice. And very few question its relevance. But is this framework sensible? Is this creating the highest level of good (the ethic) for the environment and the community? If every corporation applies this framework, will we be able to reverse global warming? Will we have reached our ultimate "big hairy audacious goal" (BHAG) – not that of the company's, but that of the whole world's?

I believe that our corporations have become myopic and unable to perceive and conceive

> For many businesses, sustainability means doing "less bad." Less bad is still bad. From my perspective, sustainable development will involve fighting this existing reality and making these models obsolete.

the broader effect of their businesses on earth. Let's begin by asking these basic questions: What is the comprehensive "effect" of our businesses on the earth? What should be our practical strategy toward sustainable development? Think about the number of the big corporations who report great environmental savings statements but never take into consideration the effect their product has on the environment or on the consumer. There are a number of big companies that produce unhealthy food products but report great reductions in energy and emissions, but lack the transparency to report their indirect harmful effects. There are also companies that produce household cleaning products that have unsafe chemicals; they have great environmental savings report, but ... again – you get the point.

I think there is a basic flaw in how we have framed our business concepts and their impact on sustainability. I want to say that the present "sustainability" frameworks are not enough. When a company starts to look at the effect they are generating anywhere along their value chain, they could begin to ask different questions – questions that are not just about reducing that effect, but of designing an effect that both delivers a higher level of value in the product while benefiting the earth and society.

At Seventh Generation we were motivated to appreciate the value of what the "Great Law" of the Iroquois Nation meant: *In our every deliberation we will consider the impact of every decision on the next seven generations*. The Native Americans were not about doing "less bad"; they were about building wealth back into their systems so that the future people were able to flourish. The Great Law is aspirational. It is an ideal state and I think it's a robust design principle. It is not yet a framework to convert the ideal to practical business strategy.

> The concept of sustainability appeals to me because it's not about doing "less bad," but about building wealth back into our systems so that the future people are able to flourish.

A cross-disciplinary design team can help create a process to align everyone in an organization toward building wealth into the system inspired by the Great Law. The key point is that this principle eliminates the fragmenting of the corporate triple bottom line – that is, human capital, natural capital, and profit. As long as you separate the sustainability metrics from financial metrics, you do not free yourself from the doing "less bad" framework. This is what we did at Seventh Generation: We put together a design team. We then interviewed over half of the company staff. We wanted to know not only what our employees thought about the company, but also what they thought the company could really become if it applied the Great Law. The interviews showed us what was working and what was not, and where the bottlenecks and new opportunities were. We put this information into a framework and called it the "promise framework."

The Business of Sustainability: A Different Design Question

The promise framework was structured as a hologram – not a linear construct – where a change in any part of the system affects the entire system. That is, we focused not on an individual element but on the relationships between the elements. The more folks we inspired, the more opportunity we had to sell our products, thus creating greater profitability potential and creating more good in the systems we influenced. We wanted to make "green" the "new normal" – in our view, "green" was building community wealth. Ultimately, we could see that the more profitable we became, the greater the effect we could have on the bigger system. Thus, our financial wealth was also creating social and environmental wealth. We obviously had to also create a robust product line that helped in our mission – after all, you can't be selling large jugs of soap water and say you are creating wealth in the bigger system.

> It is all about being relevant. The present educational system is not keeping up with the times. We need an efficient education system to provide new design tools for students and to inspire them to tackle Buckminster Fuller's question: "What would it take to 'make the world work' – to provide food, shelter and energy for 100 % of humanity?"

The promise framework was focused on redesigning the entire system and better understanding the broader effects of the business. You cannot be truly sustainable unless you look at the comprehensive effect of your business. And we need to shift from a doing "less bad" framework to a mentality of building wealth into our system. The beauty of the process was that we engaged the entire organization in our design process. Everyone participated in developing the promise framework. We all could see how the promise framework became relevant for all the projects in the company, from product design to marketing, and from our sales team to our retail partners. We could all appreciate the fact that the Great Law was a driving force and inspiration in our promise framework. We were confident that we would create a company committed to sustainability that would impact the next seven generations.

Three observations occurred during this process: First, innovation is democratic; it is also a spiritual phenomenon. To begin to see the invisible, to work together to co-create what is truly sustainable is a dynamic and deeply engaging process into the depths of our humanity – something rarely experienced in a cold corporate environment. Second, prototyping is an art. *No* prototype ever works in the beginning. Failure is wisdom: Fail hard, fail fast, keep holding the form that is trying to come in – it's not always quickly translatable into the finite world. Third, there is no "no" in the universe – no matter how many times you think it is not going to work, know you have not yet found the right question to lead you into the design process.

I know each company designs its own framework and set of objectives and metrics for success. Seventh Generation's was a holographic design framework where every decision can be designed for creating higher ethic, more good, and greater wealth in the system while still being a profitable business – with a focus on the next seven generations.

***

Gregor Barnum *worked for a number of years as director of Corporate Consciousness at Seventh Generation, Inc. His work focused on integrating sustainability, systems thinking, and innovation into the company. Previously, Barnum served as the director of Operations and Business Development for o.s.Earth, Inc., a consulting firm. He is working on a startup venture called CommonWise. He earned his master's degree from the Yale Divinity School with a focus on ethics.*

Chapter 17
Sustaining Population Health

Jacqueline Sherris

Consider a woman I'll call Maria, who lives in the Andean highlands of Peru. Maria is in her 30s, has two daughters (ages 8 and 12), a high school education, and works hard with her husband Benito to build a better life for her family. The family lives in a two-room home and farms nearby land, growing potatoes, quinoa, and other staples. In addition they own a cow and a few pigs and chickens; Maria weaves traditional Andean fabric and sells these to tourists; Benito occasionally works on construction crews in nearby towns; and recently they have earned some additional money hosting students and volunteers eager to learn more about Andean traditional culture and help in the small village school. Through hard work and ingenuity, Maria and Benito have been able to send their daughters to a better school in a nearby town, begin building another room to their home, and recently added a modern toilet to their home. The seeds of a significant success story are being sewn.

But last year Maria was worried. She had been having pelvic discomfort and vaginal discharge for several months, which she knew is not normal. There is no health facility in her village, though a visiting nurse comes occasionally. When the nurse visited, Maria talked with her about her symptoms and received a few weeks' worth of antibiotics, but these did not help. She knew there was a clinic in Cusco providing services for "women's problems" but was afraid to go by herself, fearing that as a poor, indigenous woman, she may not be understood or respected. Finally, though, she attended the clinic, accompanied by one of the European students who had befriended her and served as her advocate, and there she was evaluated for cervical cancer. She was very frightened; a woman in her village had problems similar to Maria's a few years ago, and she died, in pain and alone.

Maria's fear was well founded. Cervical cancer is a major killer of women worldwide, but particularly in developing countries – where some

85% of the quarter million deaths from cervical cancer each year occur. And in those countries, cervical cancer disproportionally affects the poorest women with the least access to health and development services. The tragedy of these numbers is that cervical cancer is almost completely preventable. Screening to identify early precancerous cervical lesions, and then simple treatment of these lesions to eradicate the precancerous cells, is extremely effective in preventing the disease – and the deaths associated with it. And vaccines that prevent infection with key cancer-causing viruses are now available; these vaccines offer the opportunity to prevent a significant proportion of future cancers among girls who receive them.

> Sustainability means that communities and systems integrate a new, proven health approach into existing systems for lasting impact.

But like Maria, women and their daughters in poor settings have not had access to these services. Vaccines are still new and relatively expensive. Pap screening programs – which have worked so well in Western countries – have not been sustainable at scale in the vast majority of developing country settings for a range of technical, systems, and infrastructure reasons. Further, political support for cervical cancer prevention has – until recently – been weak, partly because the problem was viewed as too complicated and expensive to address and – sadly – partly because it is a problem of poor women, who have little power in many settings.

This grim situation is now poised to change. A number of international organizations, in collaboration with local governments and champions, have been working to make life-saving services available to women worldwide. They have been tackling the problem from various angles, with an aim to establish feasible and effective cervical cancer screening programs (with new health products that can be provided by nurses and other non-physicians), and later HPV vaccination programs that could be integrated into existing systems; would be affordable and acceptable to women, communities, and providers; and could be demonstrated to reduce the growing toll of suffering and death associated with cervical cancer.

What are some key lessons learned from this effort to influence sustainable change in addressing a major women's health problem?

First, start with a clear goal that considers long-term sustainability and impact. The overall goal of the range of groups working on cervical cancer prevention was to reduce

> Given the health and development challenges we face worldwide, and the limited resources available to address them, we must spend our energies on efforts that have the most chance of producing a lasting impact.

the high death rate from cervical cancer among developing country women. While some groups were working to develop new tests and vaccines, others to educate women and providers, and still others to develop information systems to track and evaluate the impact of new programs, all had the goal of reducing cervical cancer death rates at their core.

Second, develop strong partners and champions at all levels, and listen to their needs and concerns. Making a sustainable difference is not something that can be done alone. Researchers, international health experts, community-based organizations, medical societies, private sector health product developers, governments, and others need to find ways to work together to achieve impact. And most importantly, keep the needs of the ultimate beneficiaries – in this case women like Maria, their families, and their health providers – at the core of planning and strategy.

Third, consider solutions to health challenges from a range of different perspectives. Researchers will view an issue from the perspective of questions to answer and information to be gained. Health implementers will focus on what an existing health system can absorb – that is, how products will be transported, how services will be delivered, and how different levels of health service will be linked. Private companies will consider the investment required for developing, producing, and distributing new health products. And women and their families will determine whether to seek care based on their fear about and understanding of an illness, their trust in a particular medicine or test, their belief in the safety of a health treatment, and – ultimately – how they are treated at the time they receive a health service. All of these elements are important, and considering them all, and designing health technologies and systems that address them, will maximize the chances of success.

Fourth, consider how sustainability in the focus health area will support the provision of other health services. Women who come to a clinic for a preventive service like cervical cancer screening often have other health questions and needs. They may, for instance, have gynecological complaints related to frequent childbearing or menopause, concerns about other cancers, risks related to chronic diseases like hypertension, and others. Girls who are eligible to receive HPV vaccines also likely have a range of health needs that are not being met, including booster doses of other vaccines, nutritional supplementation, information about puberty and menstrual hygiene, among others. Considering how to offer integrated services that meet these additional needs can increase

> Sustainability and impact are intertwined. If we are able to bring about important health and development changes that are sustained, we will improve people's health and well-being. That is the goal.

the acceptability and use of services, influence sustainability, and broaden health impact.

And Maria? After she finally found a clinic that could evaluate her condition, she was indeed diagnosed with cervical cancer, but thankfully at an early stage. And the clinic she visited had an association with a nonprofit organization that brought volunteer U.S. surgeons to the region to provide services not available locally. A surgeon removed Maria's tumor and her uterus, and she has an excellent chance of long-term survival without radiation or other treatments – which are not available anywhere near her home. The experience was a setback for her family, however. Maria could not work for a number of weeks after the surgery, incurred expenses for various medications and related costs, and did not feel like herself for several months. Benito had to be at home more, and money was tight for the family, who had to sell several of their animals to make ends meet. A year later, however, Maria is well, and the family is back to working hard together to build their future. A delay in her diagnosis – the more common situation in most poor settings – would have produced a much grimmer outcome.

There is still much to be done to bring the knowledge and experience we have in combating cervical cancer to build sustainable programs in poor countries. But the sustainability strategies described above bring hope for the future. Peru, for example, is embarking on a path to increase access to both screening and HPV vaccine services, in partnership with a number of external agencies, and building on the experience of pilot programs in several regions. Local experts are considering integrated solutions that build on existing services and are ensuring that client and provider needs are considered in plans for scaling up services. We can be hopeful that such efforts mean that Maria's daughters don't have to go through the frightening health crisis she faced, or worse, die from a preventable cancer in the prime of their lives.

***

Jacqueline Sherris *is Vice President of Global Programs at PATH. Sherris has more than 20 years of experience in public health. She has previously served as PATH's program leader for the Reproductive Health Strategic Program, through which she led and expanded PATH's cervical cancer prevention work, including efforts to increase access to HPV vaccines in developing countries. She is an affiliate faculty member at the University of Washington's School of Public Health. Sherris received her M.S. in biology and Ph.D. in science education from Purdue University.*

Chapter 18
The Struggle to Make Sustainable Change in Global Health

Laurie Garrett and Zoe Liberman

In light of the current global economic crisis, the concept of promoting sustainability as part of public health efforts has reemerged as a major focus of the global health community. Historically, however, health efforts have not been focused on sustainability. Rather, efforts have focused on disease-specific intervention programs. These programs do not support systematic change to healthcare delivery systems and in many instances have taken away medical personnel from struggling local health systems to administer disease-specific short-term interventions. Aid for global health is donated by many state and private actors in a largely uncoordinated manner, directed overwhelmingly to specific high-profile diseases – rather than toward the general health of the public. Uncoordinated aid makes things worse on the ground. Often sustainability is mentioned in passing as a project goal without the discussion of specific endpoints and strategies, but in order to achieve true long-term change, clear targets need to be at the center of any effort. Charity is easy; transformation, however difficult, is the far preferable goal.

> In the context of global health, sustainable development means meaningful improvements in the health of community members that also allows for the positive impact of interventions to continue over time. These goals cannot be achieved without support from the communities in which programs occur.

Let's consider the example of primary care. According to the World Health Organization, there's an acute healthcare workforce deficit worldwide but felt deeply in the least developed countries. Given the tremendous time and investment required to fill the necessary ranks of healthcare providers with physicians and registered nurses, near-term targets for sustained primary care delivery and basic public health must focus on paramedical personnel: midwives, community health workers, and local educators. The paucity of tertiary, even secondary, care personnel will remain a desperate feature of medicine and essential public health for at least two generations, persisting well into the 21st century.

As a thought experiment, one of us (Laurie Garrett) conceived "Doc-in-a-Box" – a prototype of a primary care delivery system for the prevention and treatment of diseases. This framework proposes transforming standardized steel shipping containers, thousands of which are abandoned every year, into transportable, cheap primary health clinics for use in developing countries. A network of converted container-clinics is achieved through franchising. The Doc-in-a-Box delivery system would address the issue of workforce shortages while ensuring ownership by locally trained healthcare franchisees that are based in the communities they serve.

Shipping containers are durable structures manufactured according to universal standardized specifications designed to allow transport practically anywhere via ships, railroads, and trucks. Because of global trade imbalances, used containers are piling up at ports worldwide, abandoned for scrap. A Doc-in-a-Box is a container that is wired for electricity and has a water filtration system, a roofing system equipped with louvers for protection during inclement weather, a newly tiled floor, and conventional doors and windows. It is estimated that large numbers of Doc-in-a-Boxes could be produced and delivered for less than $10,000 each. Staffed by trained community-based health workers, the boxes would be designed for the prevention, diagnosis, and treatment of many major infectious diseases and basic mother-and-child primary care. Each would be linked to a central hub via wireless communications, with its performance and inventory needs monitored by nurses and doctors.

Doc-in-a-Boxes can be operated under a franchise model, with the health workers directly invested in their franchises, charging for their services on a fee, voucher, or insurance basis, and realizing profits based on the volume and quality of their operations. Franchises could be located in areas that are underserved by health clinics and hospitals, thus extending health-care opportunities without generating competitive pressure for existing facilities. On a global

> Sustainability is appealing because it causes health efforts to have clear targets and endpoints – be invested long term in infrastructure and personnel training, and promote the empowerment of local recipients to increasingly take control of the effort and acquire appropriate funding, ideally through self-generated support.

scale, with tens of thousands of Doc-in-a-Boxes in place, a de facto system would exist for tracking epidemiological and disease trends, alerting authorities to urgent outbreak threats, and disseminating urgent care or vaccination in challenging circumstances.

In 2007, Rennselaer Polytechnic Institute's School of Architecture, working with the nonprofit BoxWorks, took on Garrett's Doc-in-a-Box challenge, converting a shipping container into a clinic for use in Haiti. Two years later, Elizabeth Sheehan, Founder and Executive Director of Containers2Clinics (C2C), was inspired by this idea to start an organization based on the Doc-in-a-Box. On June 14, 2010, the first 20-ft container destined for C2C's use arrived in Port-au-Prince. C2C's clinic provides a secure, semisterile, and resourced space for perinatal and pediatric care on the campus of Grace Children's Hospital. AmeriCares is partnering with C2C to deliver the pharmaceuticals used by the clinic as well as a supply to augment the hospital's existing inventory. AmeriCares staff is assisting them with project planning and logistics. By the spring of 2011, C2C's second container clinic was transported to Port-au-Prince, Haiti, and currently both are functioning in a country where there is a severe lack of access to primary care services. At this time, the C2C experiment is more about the converted boxes as delivery systems than the infrastructure of franchise-operated community health centers.

Doc-in-a-Box and the franchise model are concepts for ensuring sustainability in global health. Together they can help reach the target of significantly increasing access to primary healthcare in the developing world. VisionSpring offers a far more significantly realized franchise system for health, giving community-based women the skills and tools to test vision and sell reading glasses. In Bangladesh, BRAC, the world's largest health NGO, draws on a variety of innovative financing models to make the provision of health-related services profitable for minimally trained, otherwise poor women.

In coming decades, the most significant achievements in the provision of public health and basic primary care for the poor and remotely located populations of the world will be met through innovative financing schemes that realize the inherent entrepreneurial potentials of the poor and offer new payment schemes for health. In 2010, the WHO declared the global need for universal health coverage, recognizing that sustained improvements in health require insurance, voucher, and/or other financial mechanisms that

> Sustained global health programs can only be reached through improving health access and through community ownership – the necessary component for successful global health efforts.

give individuals purchasing power. Coupled with franchise models for basic health delivery, purchasing power could be the beginning of a sustained revolution in health the world over.

***

Laurie Garrett *is the senior fellow for global health at the Council on Foreign Relations in New York. Garrett is the only writer ever to have been awarded all three of the Big "Ps" of journalism: the Peabody, the Polk, and the Pulitzer prizes. Garrett is also the best-selling author of* The Coming Plague *and* Betrayal of Trust. *Her recent e-book is* I Heard the Sirens Scream: How Americans Responded to the 911 and Anthrax Attacks.

Zoe Liberman *is a research associate for the global health program at the Council on Foreign Relations. She received an M.P.H. in environmental health policy from Columbia University and bachelor degrees in international and global studies and biology from Brandeis University. She has worked in East Africa, South Asia, and the Middle East on global health and development projects.*

Chapter 19
Economic Growth and Sustainability Rooted in Financial Literacy

John Hope Bryant

Financial literacy – the understanding of the language of money – is absolutely essential to living a successful and independent life, promoting economic growth, and sustaining it.

That is what we are all learning as a result of a global economic crisis. It is what comes to be obvious when you recognize that the U.S. economy, the largest in the world, is 70% driven by consumer spending and that the U.S. consumer represents 25% of the world economy.

The capital markets may be driven by what is called the "big money," but our GDP (gross domestic product) is driven by consumer spending, and that means you and me. If you thought the banking crisis was bad, wait until you experience a full-blown crisis of consumer confidence. If American consumers zips their wallets shut, we are all in for a mountain of economic pain.

> Financial literacy – the understanding of the language of money – is an integral part of sustainable development. It is absolutely essential to living a successful and independent life, promoting economic growth, and sustaining it.

The reality is that our economy, here and around the world, is driven by a combination of (rational) leverage and confidence. Well, we have too much leverage (debt) at every level – from the household to Wall Street and capital markets around the world, to government, to individual households in developed countries – and too little confidence. When the consumer is overconfident and lacks financial literacy, you end up with an overleveraged economy and a credit bubble that pops, taking real and perceived value alike along with it.

Money is personal, emotional, psychological, and is tied to culture, habits, and self-esteem. If you have a low opinion of yourself, often spending money seems to make you feel better. Or worse, "trying to keep up with the Jones" as it is referred to, is an issue of identity – valuing oneself as a comparison with others – and self-esteem. We need to recognize all of this as we seek to teach anyone about financial literacy.

Financial literacy as currently articulated and taught is certainly better than nothing, but it is not going to be very effective as a meaningful and practical tool in one's economic toolbox in the 21st century, be they struggling families we are trying to help and empower, today's youth whose future we want to help shape and mold for tomorrow, the middle class that now sees how financial illiteracy impacts the quality of their lives, or policy makers to market participants.

The future of financial literacy must be "aspirationally relevant" and, I believe, rooted in the following eight practical principles:

1. *It must be aspirational.* No one wants a mortgage loan or a bond (if you live in a country such as South Africa); they want to become a homeowner. If you want to put a child to sleep, offer him or her a course in financial literacy. Respectfully, no one cares about an education in and of itself, other than an academic. Even education must relate to one's aspirations in life, in order to have relevance in their life. We must connect financial literacy to the hopes, aspirations, and dreams of people.

2. *It must recognize the very personal and emotional role that money plays in people's lives.* Teachers, volunteers, and advocates promoting financial literacy should not approach financial literacy education as if it is a simple math equation. It is not math. Money is highly personal and emotional – math is not. No one is ashamed to admit they do not understand calculus, but most people around the world are ashamed to admit they don't understand money, or their own household finances. Most domestic disputes are about money.

3. *It must be taught differently.* Traditional education around the world is based on a 20th-century agrarian model and centered around a teacher in a classroom. The only real way to reach and teach people about financial literacy in the 21st century will be through a strategy of engagement and empowerment. Really connect with those you want to reach and teach, by being vulnerable, sharing your own personal story and the challenges and disappointments you have

> The question becomes how do we move a level of individual prosperity for people at all levels, and sustain the momentum, without engaging in a wholesale giveaway program – which, by the way, never works in the long term?

met along the way in your own life. Tell them they can make a mistake, and not be a mistake. Understand that while you have been broke financially in your life, you are not poor, and neither are they. Make the experience interactive and empower those you speak to.

4. *It must be seen in the broader context of real-life experience in order to have power.* In the 21st century, financial literacy must become a localized civil right, or the first in a series of empowerment-based global silver rights, in order to have power. The 20th century was dominated by the global discussion of democracy and freedom and gave birth to civil rights movements from America to India to South Africa. In the 21st century, we increasingly live in a global free enterprise democracy, but billions of people who now have the right to vote do not have a basic bank account, an understanding of how the system works (a.k.a. financial literacy), or basic access to the levers of free enterprise, capitalism, or entrepreneurship. Without these new tools for the 21st century, individuals are not free.

5. *It must lead to an increase in global stakeholders and local taxpayers.* Financial literacy should lead to the emergence of a new stakeholder class of global depositors, new clients and customers, and owners and stakeholders, and, ultimately, new taxpayers too. Poor people cannot hire anyone, as a friend once told me. A robust and growing tax base is the only way to raise democracies, ensure safe streets, provide for a progressive social safety net, improve the standards of living, and stabilize communities long term.

6. *It must ultimately make a business case.* Financial literacy should help to make a business case for the future. If we can find a way to bring four billion of the poor into the financial mainstream, to move individuals from the working class into the middle class, and help the middle class to grow assets and over time to remain middle class, and to pass down those assets from generation to generation, everyone wins. Prosperity is the ultimate partner to peace.

7. *It must be rooted in dignity and self-respect.* The civil rights movement was not about a black man sitting next to a white man at a lunch counter in the southern states of the U.S. It was about respect, dignity, and hope. It was about empowering people to participate in the system, to see their stake in that same system, and to build a society where everyone has an opportunity to live out their

> As we move to restructure our global economy, we need to make sure that the conversation has a long-range vision, is based on a love leadership model, and that we are preparing and empowering people to participate and to become legitimate stakeholders in the system.

dreams in relative peace, security, and a sense of shared prosperity. The goal of financial literacy, if it is to have a broader relevance and a future, must be the same. It is the door that allows people to live out their dreams, with dignity and respect, in a world seemingly driven by money.

8. *It must be based on hope.* Financial literacy allows individuals, from urban cities to the most rural parts of our world, and from the minority poor in America's inner cities and Europe's poor suburbs, such as those struggling with the basics of life just outside the bright lights of Paris to the emerging economies of China, to learn to do for themselves. It allows them to place their foot on the first step of the prosperity ladder in life. It gives them a hope for the future based on love and not fear. It gives them a voice in the global conversation called prosperity.

The question becomes how do we move a level of individual prosperity for people at all levels, and sustain the momentum, without engaging in a wholesale giveaway program – which, by the way, never works in the long term?

A simple answer is a global Silver Rights Movement, where capitalism and free enterprise will overcome the obstacles of being poor.

At Operation HOPE, it starts with giving individuals the tools to understand self-sufficiency and choices. It starts with financial literacy as the door into a much broader and deeper conversation around opportunity. HOPE has served more than 1.5 million people this way since its inception in 1992. In a capitalist society and a system of free enterprise, financial literacy and a basic understanding of the language of money are the very "first silver right."

> As we move to restructure our global economy, we need to make sure that the conversation has a long-range vision, is based on a love leadership model, and that we are preparing and empowering people to participate and to become legitimate stakeholders in the system.

We must make capitalism and free enterprise relevant to the poor, and finally work for the poor. We must launch and sustain a global Silver Rights Movement.

We are all in this together; failure is not an option.

***

John Hope Bryant *is the founder, chairman, and chief executive officer of Operation HOPE, chairman of the U.S. President's Subcommittee for the Underserved & Community Empowerment, and a member of the President's Advisory Council on Financial Capability. Bryant is also a financial literacy expert for the Global Agenda Council of the World Economic Forum and innovator behind the global Silver Rights Movement.*

Chapter 20
Approaching the Future with Optimism

Robyn Beavers

The relevance of sustainability first appeared to me during my civil engineering studies at Stanford University. In the early 2000s, concepts such as replacing cement additives with fly ash or the designing of energy-efficient air condition systems were not yet questions included in my weekly problem sets. But luckily, at the time, there were easy ways to self-supplement by being proactive about the information around me. I devoured the now classic books *Natural Capitalism* and *Cradle to Cradle*. Visiting local Bay Area construction projects became my hobby. I volunteered at organizations such as Habitat for Humanity or GRID Alternatives so that I could build things with my hands to learn why and how things were built because maybe there was actual logic behind it all. And I followed and supported a new organization called the U.S. Green Building Council, which was so unabashedly challenging the entrenched building codes that dictate the future of the built environment. All of these influences and change points combined to paint a very clear picture: The built environment needs a major upgrade and it would take a vast web of decisions and actions to deliver the results we needed. It is a fascinating challenge that still inspires my career today.

> I consider sustainability to be a description of any entity that can live, grow, and prosper in a closed-loop system. The "entity" can be an economic process, an organization of people, a commercial or residential building, a vehicle, a consumer product, or a community.

Building upon my training as an engineer, I have managed to spend the last decade trying to bring to life the concept of sustainability as it applies to human interaction with the built and natural environment. At the

Stanford School of Engineering 10 years ago, it meant forging links between the best principles of engineering and the readiness of society for innovative economic changes. At Google six years ago, it meant transforming a high level of corporate commitment to sustainability into action by implementing a broad array of tangible changes from renewable energy to recycling to transportation. At the U.S. Department of Energy two years ago, it involved being part of the investment of an unprecedented amount of public money in renewable energy. At the Stanford Graduate School of Business, it meant leading a team of future business leaders in an effort to define the future impact of the water-energy nexus. And now at Vestas, it means helping to communicate the economic and environmental benefits of sustainability that wind power delivers on a global basis.

Although it might seem the tangible benefits of sustainability would be compelling to most anyone, challenges to its implementation can come from many directions and can quickly become complicated. At Google, for example, our first big investment in sustainability was an on-site solar power project that at the time was the largest photovoltaic system implemented by a corporation in the world. The objective was to install enough solar panels to provide roughly 30% of the electricity required at the Googleplex needed during peak demand. These 1.6 MW's worth of solar panels were not just about putting panels on a roof, because besides the eight needed roof tops, the panels also covered two systems of carports that were constructed just for this project and that covered the main parking lots of the campus. The scale and diversity of the project required that we involved architects, arborists, several different teams from Google, and city planning officials in the design of the installation. The solar carports were my biggest positive surprise in that their design was attractive, eye-catching, and actually added to the functional and architectural value of the office complex. Functions that were integrated into the solar carports included charging stations, WiFi antennas, and water connections. Photographs of the sea of rooftop panels and solar carports were featured in many magazines, ranging from *Fortune* to *IEEE Spectrum*. Most importantly, while the investment required several millions of dollars in capital costs, our employees were inspired and motivated by the project, the delivered savings reduced our monthly electricity bills by 90% during the sunniest months, and we estimated that the investment paid for itself within 2014 will be 7 years from switching the solar array on for the first time.

While the project was a success, the complexity and involvement of diverse stakeholders presented challenges that threatened its execution. I initially guessed that my biggest challenge would be to receive approval from the Google executive team. But the corporate leaders were more than happy to support the project, especially since the financials penciled out in

Approaching the Future with Optimism

> My belief in sustainability has offered me a life and career full of opportunity and intellectual challenges that allow me to contribute to effort that deliver concrete, positive results. I believe that the logic and essence of sustainability will continue to serve as a strong guide for my future.

a compelling way and they knew that the project could make an impact on the renewable energy market. Instead, one of the biggest challenges was to find a firm to design and build the project. At the time, no firm had ever built a project this big (except for maybe a one-off field installation in Germany) and we were not even sure we could source the nearly 10,000 photovoltaic panels it would require since the supply chain had not quite caught up to the growing demand. It is worth noting that in today's market there are plenty of firms to build a project this size, while some might even consider it too small.

However, the biggest challenge encountered was the resistance that came from the relevant local government and civic organizations. Of course, since no company had ever before planned such a large solar power system for its own use, there were zoning, building code, and public impact questions that had never been asked before. However, once these rational questions were answered, there was a frenzy of questions that crossed the line into a territory that one might consider less rational, trivial, or unrelated to the actual project. These questions were generated primarily because we were changing something: the location of a shrub, the size of a shadow, the weight on a roof, the appearance of a parking lot, etc. Ultimately, one "big" decision to make a change for sustainability (renewable energy) translated into hundreds of "small" decisions about small tangible and intangible details. We worked through these questions one by one but at the cost of more time and money.

The good news was that the installation was successful and the objectives to generate on-site power from a renewable source were met. We also learned several lessons about how to plan for and manage the inevitable translation of "big" decisions into many "small" decisions necessitated by the seemingly preternatural resistance to change no matter what the reason that society today has.

Today, I work for a company that is entirely dependent on the business of sustainability, so these types of challenges are faced in all areas of operations. At Vestas, a global leader in the wind power generation equipment industry, we approach the challenge of one big decision leading to many small decisions with our customers by addressing the sustainability benefits of all decisions big or small that are related to wind. At the big decision level, wind is the most compelling financially of all the renewable energy sources and likely has the fewest issues at the small decision level. However, the small decisions can be different. For example, solar is a horizontal renewable energy source and the small decisions tend to revolve around what is being covered up and what types of change will result. Wind is a vertical renewable energy source and the small decisions tend to revolve around what any tall structure might face and what types of change will result. In general, the big sustainability decisions for renewable energy create big positive changes; the majority of the resulting small decisions can result in positive changes as well if they are understood in advance and taken as part of the total decision-making process. At Vestas, we communicate through corporate and industry branding activities to our customers and the markets that our customers serve about the positive value and sustainability benefits of the big and small decisions that are part of the wind power solution.

As a working resident of Denmark, I sometimes feel like I am living in the new energy future. For instance, one day I decided to go and test-drive the new Renault electric car that was being introduced to Denmark. I hopped on my bike and waved to towering wind turbines offshore as I rode along the coast in my well-designed bike lane toward the Renault showroom. Not many countries in the world can offer a fossil fuel–free day like this one. Denmark has worked very hard to implement the right policies and public awareness campaigns, and as a result the average Danish citizen has embraced sustainability in his or her daily life and business approaches, and for the future of the national economy.

> Sustainability should recover from its overuse as a buzzword or slogan and return to it its original form as a powerful term for a concept and principle that can govern sound decision making, foster innovation, and continually prepare our society for the future. It needs to continue to symbolize opportunity.

Five years ago, it was clear that Europe was leading the world in the creation and use of aggressive policies that provided competitive pricing, tax incentives, and government initiatives for the acceleration of renewable energy deployment. However, since 2008, there have been economic and political forces that are not only challenging the European leadership in sustainability but also generating a wide range of new types and sizes of incentives. China has dramatically increased the size of its incentives and direct investment in both the supply and demand sides of renewable energy, with a focus on the cost reduction of proven technologies through economies of scale. Direct investment by the U.S. government in the supply side of renewable energy has surged dramatically, especially with a focus on new innovative but high-risk technologies. However, the global financial industry economic meltdown has generated new volatility in the availability of private and public funds as well as in the political will for the stimulation of and investment in renewable energy and other sustainability sectors. At the same time, China's focus on lowering the cost of existing renewable energy technologies is causing competitive ripples across the U.S. (with the bankruptcy of several heavily government-supported high-tech energy startups) and Europe (with new Chinese competitors challenging the growth of leading European renewable energy companies). I believe that this new volatility is not so much a cause for alarm but a new reason to refocus on the most important aspects of sustainability policy.

So, yes, I am an optimist. But this optimism is rooted in idealism based on reality. If we want sustainability to become an even stronger guiding economic philosophy, then we need to open our eyes to see things as a connected set of actions and processes; we need to want to make them operate better; and we must strive to maximize the use of the renewable and cyclical phenomena of nature. On a more tangible level, the results and facts show that sustainability is evolving from a slogan to a detailed and defined set of principles and practices that are often more complicated than they need to be because of the initial conditions from which they are starting. If you had asked me 5 years ago whether I would be jumping on phone calls with people representing five continents to talk about sustainability because it was our job, not just because it was our passion, I probably would have laughed out loud. But now, these phone calls, these strategic decisions, these careers in sustainability are part of business as usual all over the world. And it is clear to me that for the concept of sustainability actually to be sustainable, it has to be brought to life in a way that is economically compelling as well as intellectually appealing.

The journey to a fully sustainable society and economy is a long one that is just beginning. Taking the longer view of what can be done requires an optimism that it can succeed. It also requires the cold analytical tools to identify and evaluate the most compelling benefits, the patience to see it through new challenges, and the commitment to innovation to rapidly take advantage of new opportunities as they present themselves. It also requires that we have a clear and simple message about the goal so that the decisions at any level, big or small, will be made by closing all the loops.

***

Robyn Beavers *was the founding member of Google's Green Business & Operations strategy team. She served as an ARRA Fellow for the Department of Energy in 2009. In 2010 she was brought in by Vestas Wind Systems to help pioneer the creation of an international NGO called WindMade. She currently works for Dean Kamen as head of commercialization for his water and power technologies. She holds both a B.S. in civil engineering and an M.B.A. from Stanford University.*

Chapter 21
A Decent Place to Live

Jonathan Reckford

Before Habitat for Humanity began work in the village of Varjada in northeast Brazil, life was not easy. Each day brought the same regimen – and the same struggles. During the long, dry season – which lasts much of the year – the women would spend half the day walking 6 km (almost 4 miles) round trip to collect water from the community tank. Many members of the community also were plagued by Chagas disease, which is caused by beetles, also known as kissing bugs.

Who could have imagined that concern over these tiny insects would dramatically change the lives of the entire community?

Prior to the Habitat projects, people in the village once lived in mud huts where the walls were a breeding ground for the kissing bugs. A woman "known as" Doña Tata explained that the small beetles feed on the skin at night, usually on the face, and leave behind a parasite that causes the disease. It can be years before serious symptoms occur, but those who do get sick later may suffer from deterioration to the heart muscle, which often has fatal consequences.

> Sustainability is the capacity to endure. It is the intent to meet the diverse needs of many people in existing and future communities.

Habitat for Humanity partnered with Doña Tata and others in the community to build basic block houses with concrete floors and proper roofing. This eliminated the environment for the kissing bugs. Then, with funds from the Methodist Church, we built large stone cisterns alongside each home and designed the roof systems to capture all the runoff in the rainy season, providing enough fresh drinking water to carry the families through the year.

The women of the community, who no longer spent half the day collecting water, developed a profitable embroidery group. Through a partnership with the national bank of Brazil and World Vision, their sewing has provided them with both income generation and socioeconomic representation. The state and local governments also have worked with residents to create a school and a healthcare center in the community.

Varjada illustrates very specifically what we see worldwide – that creating partnerships to develop affordable housing can become the catalyst for transformation and for sustainable community development.

Our vision is to create a world where everyone has a decent place to live. Through volunteer labor and donations of money and materials, Habitat builds and rehabilitates simple, decent houses alongside homeowner (partner) families. The houses are sold to partner families at no profit and financed with affordable loans. The homeowners' monthly mortgage payments become part of the Fund for Humanity that is used to build still more Habitat houses.

If we are to continue to serve more families – note that I did not say if Habitat for Humanity is to continue to exist – we must determine how we can best meet the needs of families today without compromising the future.

In an effort to make houses more affordable for individual families, a number of local Habitat groups had developed policies to significantly subsidize the cost of houses. While recognizing the spirit of generosity that gave rise to such policies, our board of directors several years ago mandated that in every region of the world in which we operate, we must assess the true cost of homes. Habitat affiliates in the U.S. and national offices worldwide were directed to identify the amount of money they were subsidizing each home they built in relation to the amount of money that they could recoup and use for future construction. The question we had to answer was difficult: With finite resources to invest, how can we maximize our impact? We are not against subsidies, but with an eye toward the future, we had to determine how we could best leverage our funds and consciously target subsidies to those with the greatest need.

Our mandate is clear: We are to seek poverty housing solutions for as many people as possible. Therefore, we had to develop systems that would ensure our ability to continue working with those we are called to serve. That challenged us to come up with simpler, more affordable housing solutions, and it meant we had to be more strategic about how we use subsidies.

But what about the things you can't plan for and couldn't possibly predict? How were we to respond to crises like the tsunami that destroyed thousands of homes in Asia, hurricanes Katrina and Rita, the earthquake in Haiti, or the failing U.S. economy? These events, which devastated entire

> Developing sustainable communities is economically, environmentally, and morally the right thing to do. It calls upon those whose welfare is at stake to help make decisions about how their lives and the lives of generations beyond can be improved.

communities, meant that we could no longer do business as usual. We had to rethink our entire approach to helping families in need of shelter.

The level of response required following the disasters, which seem to keep coming harder and faster, forever changed our idea of what was possible.

We had to find ways to engage huge numbers of volunteers, to process streams of homeowner applications, to distribute tons of materials, and to properly direct extraordinarily generous gifts. We developed strong partnerships with individuals and groups around the world, who continue to support our affordable housing efforts.

This idea of partnering with other organizations and addressing large neighborhood challenges turned out to be an important strategy in the U.S. as well. As the recession moved deeper into the lives of Americans, neighborhoods became glutted with foreclosed and vacant properties – which are toxic for surrounding areas. As their numbers kept increasing, it was obvious that local communities needed rejuvenating.

We realized that we had to make some changes. If we were going to continue engaging volunteers to provide housing solutions, we needed to encourage our affiliates to think not only about new construction, but also about repairs and weatherization. A key to our efforts would be to purchase foreclosed properties to rehab and sell at an affordable price to low-income families. This idea of neighborhood revitalization – of making a greater impact in communities – is becoming the model of how Habitat will operate going forward.

Donors like this idea of a bigger vision and of operating on a larger scale. We have adapted to the catastrophic events we never expected and have turned calamity into opportunity.

Our operating strategies may have changed, but our vision is still the same: to create a world where everyone has a decent place to live. We seek to achieve the greatest possible impact, in terms of improving the lives of families, with the resources at our disposal.

If our focus was merely on building houses, we could measure success by counting the ways we replicated the house-building model in a certain region. However, if we are assessing long-term impact, we must determine if our strategic interventions have resulted in lasting reductions in poverty housing. Put more directly, second-generation Habitat homeowners should be rare; third-generation Habitat homeowners should be unthinkable.

We want to see families continue to improve their situations and those of their children.

Our task is to help people move out of poverty and into homes and communities where they can thrive. We want to help families create neighborhoods that foster success.

When people have financial, physical, and emotional investments in their homes and their communities, the future can be exciting and the choices many. Following the devastating tornado that struck Evansville, Indiana, in 2005, the community rallied with an incredible response. The tornado that wrought such destruction also became the catalyst for amazing rebirth.

Residents decided that if they could rebuild after the violent storm, they could rebuild their center city core. Multiple nonprofit organizations and private and public partners have worked intensively with the neighborhood residents since the fall of 2008 to create a shared strategy for neighborhood improvement. Listening to people in the community was a crucial first step. Community building led by the neighborhood and supported by external partners is a centerpiece of this initiative.

Habitat Evansville has built 23 homes across the street from the new Glenwood Leadership Academy, which includes a K–8 school, a health clinic, a new gym, and a media center. A community garden has also been planted. The city government, the local housing authority, community foundations, area universities, local businesses, and the state Department of Energy are among the partners that are helping to clear out blighted properties, establish green spaces, build energy-efficient homes, increase access to healthy foods, and create retail and educational centers.

Vibrant neighborhoods impact many people and create a ripple effect of benefits. Homeowners pay taxes that support government services for a larger population. Children living in stable homes are healthier and do better in school. Adults who experience success often return to school and obtain better jobs.

Though Habitat for Humanity is best known for the images of volunteers raising the walls on a worksite, our efforts are strategically directed beyond the construction of individual houses. Habitat is committed to the larger and longer process of creating partnerships that can transform lives and communities for generations to come.

The way we go about serving families is deeply rooted in concern for the future of the environments in which we operate. We will do no harm to those we seek to serve, to the world around us, or to ourselves. We must leave the world, on balance, at least as well off as we found it.

That means we are conscious of waste, and we are quite creative in our recycling efforts. One of our affiliates set out leftover building materials with a "free" sign and made the items available to the neighbors. It worked. The amount of waste was minimal.

We are committed to green building by constructing energy-efficient homes and using materials that are least burdensome to the environment in terms of manufacturing and transportation. A homeowner family that partnered with our Tacoma-Pierce County (Washington) affiliate on a green-built home gets an annual rebate payment from the utility based on production (regardless of whether they use the power they produce or if it goes back on the grid). Their total electrical power cost for their first full year in the home was $25.92 – about 7¢ a day!

We have established a standard that by 2013, all Habitat homes in the United States will have, at the minimum, an Energy Star rating. The vast majority of our affiliates already are building houses to that level and beyond. Several are building to the highest level of LEED certification.

> In the context of Habitat for Humanity, some people limit the concept of sustainability to ideas about green building. While that is important, it is but one facet of our commitment to meet the needs of families, both now and for generations to come.

Our priorities are to provide decent homes in decent communities at a cost that families can afford. Utility and transportation expenses must be considered when determining the true cost of living in a home. Money that homeowners can save on lower utility bills is just as valuable as a no-profit mortgage.

Rooted in the commitment to demonstrate the love of Jesus Christ, Habitat for Humanity takes seriously the call to be good stewards, to responsibly manage all that God has entrusted to our care. That means that we use our resources in a way that does not result in depletion or permanent damage.

By creating sound financial policies, by helping families build a foundation for a better life, by developing community partnerships, and by committing to be good stewards, Habitat will best be able to meet the needs of families now and in the future. Focusing on sustainability on every level will allow us to further our mission to build homes, communities, and hope.

At Koinonia Farm in South Georgia where Habitat for Humanity has its roots, faithful leaders envisioned partnership housing as a demonstration plot for sharing God's love. This excerpt from a prayer offered for Archbishop Oscar Romero describes beautifully our intent:

> This is what we are about:
> We plant seeds that one day will grow.

We water seeds already planted, knowing that they hold future promise.
We lay foundations that will need further development.
We provide yeast that produces effects beyond our capabilities.
We cannot do everything and there is a sense of liberation in realizing that.
This enables us to do something, and to do it very well.
It may be incomplete, but it is a beginning, a step along the way, an opportunity for God's grace to enter and do the rest.

***

Jonathan Reckford *is CEO of Habitat for Humanity International, an ecumenical Christian housing ministry that has helped shelter more than two million people around the world. Reckford graduated from the University of North Carolina Chapel Hill and the Stanford University Graduate School of Business before spending much of his career in the for-profit sector, including executive and managerial positions at Goldman Sachs, Marriott, the Walt Disney Co., and Best Buy. He also served as executive pastor at Christ Presbyterian Church near Minneapolis, Minnesota. Reckford is author of* Creating a Habitat for Humanity: No Hands but Yours. *He is a member of the Council on Foreign Relations and serves on the boards of Interaction and the Urban Land Institute Terwilliger Center on Workforce Housing.*

Chapter 22
From Field to Market: Changing Our Focus

Gerald Steiner

Five years ago, I found myself in a vigorous debate with a group of farming, food, and environmental leaders in a small conference room. I was there by choice. In fact, it was partly my idea to bring together a visionary group across the agricultural spectrum to see if we could improve our understanding of "sustainable agriculture." We started by discussing which management practices best described sustainable agriculture:

- *No-till* systems prevent soil erosion and allow the soil to store much more water and carbon.
- *Organic* production doesn't allow synthetic inputs and mimics natural systems.
- *Locally produced food* helps consumers understand how their food is produced.
- *Biotech crops* increase yields, while enabling better weed control and fewer insecticide sprays.
- *Closed-loop* systems require very few purchased inputs and are buffered from price shocks.
- *Pesticide-free* production meets the needs of consumers who have concerns about chemicals in their food.
- *Cover crops* prevent soil erosion, add nutrients to the soil, and increase soil carbon over time.
- *Integrated pest management* (IPM) systematically minimizes pesticide applications and is the right balance between organic and conventional.
- *Profitability* supports the farmers to survive another year and plant the next crop.

I was bringing my own definition and priorities to the debate. I grew up on a dairy farm in Wisconsin and learned firsthand to appreciate the challenges farmers face. I saw my dad struggle to keep our small farm profitable. Our family had pride in caring for the land, but the notion of sustainability was never as important as getting the work done and paying the bills.

At the start of my career with Monsanto in 1982, farmers were on the verge of a crisis as debt levels increased and commodity prices crumbled. Supply management schemes to keep food cheap and farmers employed were the prevailing government policy objectives. For the first 20 years of my career in agribusiness, these conditions prevailed, with an occasional oasis of profitability on the farm.

> Our challenge is to meet the demands of today while preserving the resource base for a growing population to do the same in the future. From an agricultural standpoint, to make this happen requires thinking about more than just the quantity farmers produce. They need to be profitable. Collectively, we have an obligation to protect the environment. We need ideas to compete and new methods to collaborate. To me, achieving all of this *simultaneously* is sustainable development.

It was during this time, however, that a global economic revolution was quietly starting. The fall of the Berlin Wall, the collapse of the Soviet Union, and careful movements of the Chinese to install a more market-oriented economy were changing the world. A growing global population coupled with growing economic liberalization at a scale the world had never experienced was sowing the seeds of a demand boom for food, water, and energy. I saw evidence of this as I traveled around the world, especially in Asia.

It is now widely acknowledged the drivers have changed in agriculture. Governments are less worried now about keeping farm commodity prices profitable and more focused on bridging the gap between supply and demand. There is less debate about the food security challenge or the critical role that farmers – as a cohort of uniquely independent decision makers – will also have on the environment and global prosperity.

> Sustainability is effectively the legacy of our life's work. My company, for example, seeks to serve farmers, making them more productive, resource efficient and profitable, so they can also serve their communities. If we can do this well, it's a win–win–win and future generations will be the biggest winner of all.

Back in this small conference room 5 years ago, the discussion about sustainable agriculture started with a focus on management practices each were intent to defend. With some skillful facilitation, the discussion evolved to focus on the outcomes that we cared most about. I was shocked how quickly the group used this technique to galvanize around a definition for sustainable agriculture we could all agree with.

Once our definitions were settled, someone asked, "How are we doing?" As a group, we realized that the U.S. agricultural sector didn't really know if we were on a sustainable path toward the future. There were no systematic measurement and reporting mechanisms on the economic, environmental, and social impacts of agriculture. In the business world, we call this "not knowing where our business is." It is a completely unacceptable place to be as a manager.

We committed to take action on this issue by commissioning a study focused on a narrow set of environmental indicators for four crops produced in the U.S.: corn, cotton, soybeans, and wheat. We stayed focused on measuring outcomes and resisted the temptation to measure the adoption of various best management practices (BMPs).

> We need to generate more clarity and awareness on the critical outcomes and scale of impact required to meet present and future needs. Too often, we resort to defending specific "best practices." By staying focused on the necessary outcomes, we would empower our most creative people to develop innovative solutions.

As a starting point, the study was revealing. There has been significant progress using resources efficiently and limiting ecological impacts in some crops, but the progress was uneven. For example, over a 20-year period of time, the total greenhouse gas (GHG) emissions required to produce 1 metric ton of corn had declined by 30%. At the same time, the GHG emissions to produce 1 metric ton of wheat grew by 15%. What was driving this difference, and what could be done to manage to a different outcome? We started asking those questions within my company as well. Are there ways we could improve our products that would reduce the greenhouse gas footprint of the crops our farm customers grow?

Field to Market: The Keystone Alliance for Sustainable Agriculture started as a simple dialogue about a phrase than many in the agribusiness community felt had lost meaning. Encouraged by our early efforts to measure the true impacts of the agriculture sector, the group has grown to include nearly 50 member organizations and is developing a strategy for new sources of value creation around reaching mutually agreed-upon sustainability goals. It has quickly become the most diverse, cutting-edge organization we are a part of.

Our involvement with the start of *Field to Market* gave me the confidence to challenge our company to set measurable targets for performance improvement around the economic, environmental, and social outcomes resulting from the use of our products. Monsanto announced our Sustainable Yield Initiative (SYI) in 2008. Now we simply call it our "commitment to sustainable agriculture." It has become the vision for our company, and the strategic imperatives have permeated every part of the organization.

I think the business community has a lot to gain by becoming more transparent and acknowledging the true extent of our interdependence in the discussion about sustainable agriculture. Measuring and reporting the outcomes of our products are a great way to start that process.

***

Gerald (Jerry) Steiner *is Executive Vice President for Sustainability and Corporate Affairs at Monsanto Company. Steiner has helped establish several new pre-competitive sustainable agriculture efforts, including Field to Market: The Keystone Alliance for Sustainable Agriculture, Global Harvest Initiative and the World Economic Forum's New Vision for Agriculture project.*

Chapter 23
Joules: The Currency of Sustainability

Chandrakant Patel

During my travels in India, I often spend a great portion of my time with local micro-businesses such as roadside vendors of vegetables or local general stores. These businesses, locally called *kirana* stores, serve the local communities or "societies." In the mornings, I enjoy watching these businesses come alive. I watch the laundry man gingerly fill red hot coal into his massive clothes iron and learn from him that it is more economical to use a coal-fired iron to press clothes than an electric iron.

In the evenings, I usually saunter over to a general provision store. One such evening, I found myself in my village's general store, the size of a one-car garage in the U.S. Among many customers that evening, one in particular caught my eye. After the customary jovial greetings typical of the region and the culture, this customer handed 3 rupees – less than a dime in U.S. currency – and a small glass bottle to the shopkeeper. The shopkeeper partially filled the bottle with about 100 milliliters of a blue liquid from a large container to complete the transaction. The customer turned around to leave, paused, fiddled with some currency, and returned to purchase a 100 g of rice and similar quantities of lentils and vegetables. While I was not surprised by the small quantities of food purchases, I was perplexed by the blue liquid and asked the shopkeeper about it. The shopkeeper replied that the blue liquid was kerosene for the lamp in the family tent and the rest were supplies for the family's evening meal. The customer,

> Sustainability means minimizing the destruction of available resources and thereby preserving the finite global pool of available energy for future generations to continue to enjoy the same quality of life as the current generation.

a laborer who works in the booming sugar cane fields in the state of Gujarat, India, is one of hundreds who rely on that small shop for their fundamental provisions on a daily basis. In turn, the shopkeeper provisions his supplies based on projected needs of the clientele. The shop has a mini-refrigerator that stores 200 ml, 5-rupee, pouches of milk. The supply of milk comes from a superb model of cooperative dairies that are prevalent in the region. There is no space in the refrigerator for vegetables and other perishables. Therefore, maintaining the right balance of supplies is critical for the shopkeeper.

That evening, as I marveled at this well-provisioned supply–demand operation, scores of gleaming cars jamming the entrance to the small village caught my eye. The cars, of vivid makes and sizes, represented the phenomenal growth engine and the consumption powerhouse that is India today. The gainfully employed laborer also represented the widespread participation of the populace in the nation's growth. But therein lies the challenge: How do we maintain the quality of life of the half a billion, the likes of the laborer, in the face of the strains the societal growth is placing on the available resources? How will the shocks associated with environmental externalities impact the lives of the laborer and the vegetable vendor? Indeed, recent military, social, and political perturbations in the Middle East resulted in an increase in the cost of fuel and basic necessities.

> The concept of preserving the finite global pool of available energy is appealing to me as it enables a framework that can price a product in terms of joules of available energy destroyed across its lifetime. Quantifying and minimizing the destruction of available energy used in extraction, manufacturing, transportation, operation, and end of life is strategically important for devising products and services in a resource-constrained world. The approach also lends itself to the systematic application of basic principles of engineering.

In this context, the current global interest in sustainability presents an immense opportunity. While we have a vast portfolio of technologies that can be applied for better need-based provisioning of resources, we have yet to apply a systemic approach in devising scalable solutions. The information technology (IT) ecosystem made up of billions of handheld devices and thousands of data centers can play a key role in this transformation. As an example, the shopkeeper, who extensively relies on the wireless phone network and innovative applications of text messaging, could further reduce his sales costs with web access and IT services that enable better provisioning of his supply. The shopkeeper would like Internet access at 50 rupees, or about $1 per month. In order to facilitate this transformation – by enabling services at US $1 per month to bring billions like the shopkeeper on board – the information technology infrastructure itself must be devised and operated with the least energy and least materials. Indeed, contrary to the oft-held view of "paying more to be green," addressing sustainability with an end-to-end life-cycle perspective will lead to the lowest-cost IT services. In order to reduce the total cost of ownership by approximately one fifth from the current state and provide low-cost IT services to a broad ecosystem of clients, IT has to be sustainable before IT can be applied for sustainability.

Unfortunately, sustainability is rife with anecdotal thinking amounting to "dos" and "don'ts" lectures by pundits. It lacks an irrefutable holistic framework based on supply and demand. Recently, my team has proposed a framework built on available energy. Simply stated, the second law of thermodynamics dictates that all the energy in a given form cannot be converted to useful work. As an example, kerosene has about 45 MJ/kg of available energy. Only part of this amount can be converted to electricity; available energy (or "exergy") refers to this useful part. Once the kerosene is burned, some fraction of its available energy is destroyed – and a valuable resource is lost from a finite supply pool. Indeed, we will have to wait millions of years for dinosaurs to return our global pool of fossil-based available energy.

Therefore, we propose a framework based on available energy that minimizes the destruction of exergy and harnesses as much as possible from waste streams. Then we use a supply–demand framework that matches availability based on the needs of the user via the following principles:

- *Principles on the supply side:*
    - Minimize the available energy required to extract, manufacture, mitigate-waste, transport, operate, and reclaim components
    - Design and manage using local sources of available energy
        - To minimize the consumption of available energy in transmission and distribution, e.g., dissipation in transmission,
        - Take advantage of available energy in the waste streams, e.g., exhaust heat from an engine

- *Principles on the demand side:*
    - Minimize the consumption of available energy by provisioning resources based on the needs of the user by using flexible building blocks, pervasive sensing, communications, knowledge discovery, and policy-based control

> Sustainability has become a marketing mantra, and actions deemed "green" on the part of producers and consumers are narrow in scope and often lack validation. The most significant improvement in understanding and practicing sustainability would be to define and apply a supply–demand framework and to emphasize a curriculum steeped in fundamentals of science, engineering, sociology, and economics.

Let's examine the first principle on minimizing the available energy required in extraction, toxic waste mitigation, manufacturing, transportation, operation, and reclamation of products. This lifetime available energy approach can be built using principles of thermodynamics and can become part of the practitioners' toolkit. The use of computer-aided design and engineering (CAD and CAE) to define and analyze products is commonplace today. As an example, my daughter is learning to be a mechanical engineer and is being taught such traditional design tools. In a world where lifetime available energy consideration will drive product design, she ought to also learn the use of an available energy analyzer as part of the CAD tool to value the product in joules of available energy destroyed in its lifetime. As part of the analysis, she can also estimate the available energy that can be harvested from the waste streams, such as heat energy from furnaces used to manufacture the product. For example, classical thermodynamics dictates the amount of energy that is available from waste streams such as waste heat energy from an engine (for example, 1 J of heat energy at 500°C has approximately 0.6 J of available energy). In addition, at design time, she can perform "what-if" analysis on material choices, amount of material, supply chain options, etc., to devise a least-joules product. The key challenge in applying this principle lies in having a repository of available energy data for a variety of materials and processes. In the case of our shopkeeper, designing the least-lifetime-joules handheld phone and the data center to provide useful IT services necessitates such a repository of data, including available energy data for the extraction of various materials used in building the handheld device and the data center. Given the enormity of such a database and the need for inputs from practitioners, the IT ecosystem can once again enable us to build an open global "sustainability hub" where the world contributes this knowledge.

Thus, the joules of available energy consumed across a product's lifetime becomes the currency. Reducing available energy consumption preserves the global resources and reduces greenhouse gas emission. The application of this approach during design and build process in all ecosystems, from cars to buildings, will enable us to value and provision our critical resources judiciously. In addition, when combined with other principles of the supply–demand framework, "right provisioning" of resources can also help mitigate the inflationary effects of global growth and consumption. To be sure, there are challenges to implementing the supply–demand framework – but we can get started with existing knowledge today.

The world cannot wait.

***

Chandrakant Patel *is an HP Senior Fellow and interim director of Hewlett-Packard Laboratories. Patel has been a pioneer in microprocessor and system architectures, management of available energy as a key resource in "smart" data centers, and, most recently, application of the IT ecosystem to enable a net positive impact on the environment. Patel has also taught at Chabot College, University of California, Berkeley, Santa Clara University, and San Jose State University. Patel is an IEEE and ASME Fellow, holds a bachelor's from the University of California at Berkeley and master's from San Jose State University, and is a licensed professional mechanical engineer in the state of California.*

Chapter 24
Innovation Economics: The Race for Global Advantage

Robert Atkinson

Several years ago, I was invited to India to give a series of lectures on the innovation economy. My host transported me from the Kolkatta airport to the five-star Sonar Bangla Sheraton Hotel, located in a poor neighborhood. The next morning I asked the concierge for a place to jog and he informed me that they had a jogging path inside the walled hotel compound. When I said I would rather jog outside, he responded, "Oh no, sir. That would be dangerous." Deciding to take my chances, I exited the compound and started down a small side street. I soon jogged past a mother and her three small children sitting on the dirt floor of a small tin roof shack. About 100 yards farther, I passed a father and 10-year-old son atop a mound of rocks breaking large rocks into smaller ones with hammers. As I got back to the hotel, I was struck by the contrast. There I was in a hotel where one night's stay cost a month's wages for the average Indian and probably half a year's wages for the families I had just seen.

> Sustainable development means to me a rapid and continuous growth in the standards of living of peoples around the world, particularly citizens of developing nations.

Perhaps I should have felt guilty. After all, we are told that it's Western affluence that consigns developing nations to poverty. "Live simply so that others may simply live," read the bumper stickers plastered to countless Toyota Prius bumpers. My Indian-born parish priest summed it up one Sunday when he blamed Indian poverty on Americans' excessive materialism. But I didn't feel guilty, because American wealth doesn't come at the cost of Indian economic well-being; it actually helps them (where would India be without computers, jet aircraft, and advanced telecommunications?).

Rather, I felt distraught. At the same time I felt relief; relief that I was fortunate enough to win the global lottery and be born in the U.S. (actually Canada, but close enough), where even minimum-wage workers' lives are vastly better than the lives of the people I had just seen. Finally, I felt disbelief and outrage. How could people put up with this? Why didn't they riot? And how could India's leaders not be doing more to grow their economy?

A few days later, I gave a talk to a group of leading Indian CEOs. Afterward I chatted with several CEOs. I mentioned that during the cab ride from the airport I saw a group of about 20 workers doing road work. On the side was a pile of gravel, and each worker was going to the pile, filling a small bucket, and then walking with the bucket on his head to dump the gravel into the hole. I said, "I understand the notion of labor–capital ratios so maybe the contractor can't afford to buy earth-moving equipment, but surely he could have his workers spend a day making three wheel barrows. What they lost in the time of making the wheelbarrows they could surely make up in one day of higher productivity by just three workers with wheelbarrows do the work 20 did." I will never forget what one of the CEOs told me: "Rob, you don't understand. In India we can't afford productivity." What he meant was that the need to employ people is so great that they can't afford to boost productivity since it leads, at least in his mind, to fewer jobs.

I might have expected a liberal social activist to say this – after all, for most of them productivity is bad since it would mean fewer workers moving the gravel. But to hear it from a CEO was, to put it mildly, a shock.

It would be bad enough if this was just some unique cultural proclivity against productivity that Hindu nations have. But this view is endemic. A number of years ago, I was at a luncheon sitting next to the economic advisor to the Chilean President. I related a similar story of inefficiency in Chile and wondered why they didn't take the easy steps to be more efficient. His response was the same: "We have to create as many jobs as possible, and boosting productivity impedes that."

For all the praise it gets as a fast-growing economy, the same view is widespread in China. On a recent trip there, everywhere I looked, what was done in the U.S. by one or two workers was done in China by a multitude of workers. My hotel's front desk was staffed 24 h a day with seven or eight clerks, although I never saw more than two or three guests there. At the pool, three workers staffed the cabana, although it being December, I only saw one hearty guest braving the unheated pool. At a local deli, three people handled paying for the sandwiches: One put the sandwich in a bag, the second took the

> The idea of sustainability appeals to me because it is difficult to be fully human unless one is free of the constraints of poverty and overwork.

money, and the third put money in the register and handed the change back to the second person.

This intentional inefficiency is endemic in many developing nations. When the Kenyan government recently gave a Chinese company a waiver to import batteries without paying a mandatory 35% duty, the local Eveready factory had no choice but to install more machines to cut costs. But what is striking is that Eveready had long resisted automation in order to pad employment levels, even though it meant Kenyans paid more for batteries. This is not unusual. Many multinational firms in developing nations report doing this as a result of being under government pressure.

This brings to mind Milton Freidman's famous quote. In the 1960s on a trip to an Asian nation, he was taken to see a project where the workers were building a canal. He was struck that the workers were digging with shovels. Friedman asked why they weren't using earth-moving equipment, to which the government official proclaimed, "This is a jobs program." In turn, Friedman responded, "If you want jobs, why don't you give them spoons instead of shovels?" That really is the choice for much of the developing world: spoons or tractors. The current regime of "spoons" means consigning billions of people to grinding poverty. "Tractors" mean lifting them out of poverty.

So it is impossible for me to write about sustainability without addressing the harsh reality that for billions of people, life is barely sustainable. They work long hours, in backbreaking, soul-numbing work for almost no money. That is not sustainable, for them as individuals or for humanity as a whole. Sustainable development thus means not only lifting virtually the entire global population out of poverty, but it means building economies that are productive enough to let people work a reasonable number of hours in a year at jobs that are reasonably interesting and provide decent incomes. Sustainable development means getting all seven billion inhabitants to developed nation standards of living as quickly as possible while enabling developed nations to continue on to the next level of creating more satisfying workplaces, reducing work time, and providing increased choices of where people live, leading to the creation of more livable communities.

I would imagine that at this point many who are concerned with sustainable development are thinking that this vision is the opposite of sustainability. Isn't it, in fact, the environmental nightmare people like John Malthus have been warning about for over 200 years? Isn't sustainability supposed to be about reducing the human footprint on the biosphere? Isn't sustainability about using less, not more?

But keeping seven billion people poor in order to save the planet is not only the antithesis of sustainability, but it's not even an answer for environmental sustainability. That's because the only way to save the planet from climate change and resource-based degradation is to make people rich enough so that humankind invests much more resources into environmental and energy innovation. For it is only through innovation that we can transform our energy system from a carbon-based one to a carbon-free one. Once this is done, planetary GDP can be 5, 10, even 20 times larger, with greenhouse gas emissions a fraction of today's level.

> Environmental sustainability is really only possible through radical technological innovation, especially clean energy innovation.

This transformation to clean energy is inevitable. The only question is whether policy makers will devote the resources to support clean energy research, development, and deployment on the scale and time frame needed. But the richer the world gets, the more resources there will be for clean energy innovation and the faster we can get the job done.

Finally, let's get to the claim – the *excuse* really – that "we can't afford productivity." While it is obvious that if those Indian workers used wheelbarrows, 15 of the 20 workers who were using buckets would likely lose their jobs, the story doesn't stop there. Because the remaining workers would be more productive, they could be paid higher wages, which they would use to buy more goods and services, which some of the other 15 workers would produce. Likewise, the cost of repairing the road would be lower, so the local government would be able to cut taxes, meaning that the citizens would have more money to buy goods and services, in turn employing more people. This common-sense view is borne out by the scholarly literature, which finds that at least in the moderate term, higher productivity leads to more jobs, not fewer.

So, at the end of the day what is not sustainable is low productivity and poverty.

***

Robert Atkinson *is the founder and president of the Information Technology and Innovation Foundation, a Washington, DC-based technology policy think tank. Before coming to ITIF, Atkinson served as vice president of the Progressive Policy Institute, the first executive director of the Rhode Island Economic Policy Council, and project director at the former Congressional Office of Technology Assessment. Atkinson has testified before a number of committees in Congress and has appeared in various media outlets, including CNN, Fox News, MSNBC, NPR, and NBC Nightly News. He received his Ph.D. in city and regional planning from the University of North Carolina at Chapel Hill. He is also author of the books* Innovation Economics: The Race for Global Advantage, and Why the U.S. is Falling Behind; The Past and Future of America's Economy: Long Waves of Innovation That Power Cycles of Growth; *and the* State New Economy Index *series.*

Chapter 25

Unlocking the Energy of Business to Effect Change

Meg Crawford

Your paper coffee cup from Starbucks tells you it is made from 10% recycled postconsumer content. The care tag on your Levi's 501 jeans urges you to save energy by washing your jeans in cold water and to avoid the landfill by donating your unwanted clothing. When you do get around to washing those jeans, your Tide laundry detergent is 2–3 times more concentrated than it used to be, to help cut down on the amount of energy needed (in the form of heating the water) to clean your clothes. Clearly, companies – like Starbucks, Levi Strauss, or Procter & Gamble – are increasingly marketing to customers and stakeholders their "sustainability" credentials. Sustainability seems to mean "good" or "less bad" for the environment, or the world, and to encourage customers and the public to have a positive opinion about the product or company.

But sustainability is strategic. It is not about engendering feelings of goodwill or getting a good score in corporate reputation rankings. It is about recognizing and planning for the long-term economic, social, and political trends affecting a company's ongoing profitability. No company, organization, or society operates in a vacuum: Long-term prosperity depends upon companies that respond responsibly and effectively to the needs and conditions of the communities in which they operate, and in turn, corporations can benefit from healthier communities that help businesses to grow, innovate, and attract talent.

> Sustainability to me means using our resources wisely and efficiently today, in a way that doesn't jeopardize the well-being and prosperity of future generations.

Companies reaping the rewards of sustainability strategies don't see "sustainability" as an extraneous endeavor but rather as a set of specific considerations that are integrated into the business and guide decision making. Companies are finding that actions to address environmental challenges related to their business help to improve operational efficiencies, hedge risks associated with high and fluctuating energy costs and related greenhouse gas emissions, address customer and stakeholder concerns about the environment and public health, and respond to government policies that seek to promote clean technologies, create jobs, and ensure public well-being. Increasingly, sustainability is integrated into the practices of corporate procurement, finance, facilities, fleets, operations, supply chain, marketing, investor relations, and human resources. "Sustainable" thinking is helping companies to reduce waste, generate new business opportunities, and maintain competitiveness. In one example, an integrated, holistic approach to managing water and energy use saved tech company IBM $3 million at a single production plant through water efficiency measures, while increasing output by 33%, without incurring any capital costs. Recently, Unilever, the world's second-largest food and personal care goods company, launched its "Sustainable Living Plan" focused on mitigating the environmental and social impacts from the company's supply chain for such products as Lipton Tea and Dove, all the way down to the farms and raw materials. This commitment was made with a recognition that concerned consumers are starting to "vote with their wallets" on products that do (or do not) take into account such challenges as food shortages, malnutrition, and climate change. "Companies that do this will get a competitive advantage," says the company's CEO Paul Polman. "Those that do not will put themselves at risk."

Companies have significant resources and talent to contribute to sustainability challenges. Watch what happens when Wal-Mart, the world's largest retailer, which employs over two million people and annually procures billions of dollars in goods and services, uses its purchasing power to change behaviors among companies and individuals. In 2009, the retailer began assessing the impacts of the thousands of products on its shelves, seeking to prioritize those that are, for example, more efficient and longer-lasting to help reduce consumers' energy use and waste. Ultimately, the company will prioritize purchases from suppliers that offer innovative, affordable products that are more sustainable, creating a domino effect through purchasing decisions that will drive changes in practices among its more than 20,000 suppliers. To take an example of just one product innovation, selling only 100% concentrated liquid detergent at Wal-Mart, suppliers will help save over 400 million gallons of water, 95 million pounds of plastic, 125 million pounds of cardboard, and millions of dollars in transportation costs over 3 years.

Companies are not just balance sheets and earnings reports. They are made up of people and human energy. The companies in the Fortune 500 collectively employ nearly 25 million people worldwide. And companies of all shapes and sizes are engaging their employees to make a difference. Johnson & Johnson runs an environmental literacy program for employees to increase understanding of global environmental issues. All facilities are expected to implement a five-year literacy plan that includes a different environmental education campaign each year. In 2008, 97% of facilities ran a literacy campaign, the majority of which concerned climate change. Realizing that 70–75% of employee healthcare costs in the U.S. are attributable to lifestyle or modifiable behavior, the electric utility PG&E has launched "wellness accounts" for its nonunion employees. The company credits an employee's account when the employee completes certain activities or engages in healthy behaviors, and the account can be used to pay for eligible healthcare expenses. In 2008, Intel began including environmental factors in the calculation of corporate performance on which every employee's annual bonus is based. Three environmental performance goals are included: product energy efficiency, the company's reputation for environmental leadership, and the completion of renewable energy projects and purchases. In 2009, Intel added reducing the company's carbon footprint as a performance metric.

> Sustainability appeals to my sense of fairness and efficiency – that we should not be wasteful and that today's generation should leave the enjoyment and value of today's natural resources for the next generation.

My own interest in global social, political, and environmental issues developed from a fascination with international affairs, foreign cultures, and my innate sense of fairness. In the corporate world, fairness is a critical element of "corporate responsibility," and I have worked to advance more responsible business practices through my career and educational opportunities. My interests led me to attend Georgetown University for its strong foreign affairs curriculum, to spend time studying abroad, and to start my professional life as an intern at the International Foundation for Election Systems, where I contributed to efforts toward establishing free and fair elections in African countries. Eager for further international experience, in Paris I worked on a multiyear legal case involving the efforts of the European Bank for Reconstruction and Development to promote small business development in Russia. My ensuing professional career working in the private sector provided exposure to the inner workings of companies, and my subsequent master's degree, focused on international business and government affairs and on international business diplomacy from

Georgetown, was the closest education I could find to pursue my interest in "corporate responsibility." This additional education opened up new professional avenues, including working with companies and investors to address sustainability challenges, such as global climate change, at Ceres, a national network of investors, environmental organizations, and public interest groups.

In my current role at the Center for Climate and Energy Solutions (formerly the Pew Center on Global Climate Change), I engage in work analyzing climate-related markets and business strategy and manage the center's project on low-carbon business innovation. Much of my work involves studying the climate and energy strategies of the center's Business Environmental Leadership Council (BELC), a group of (largely Fortune 500) corporations – with combined revenues of over $2 trillion and over four million employees – that partner to address political, environmental, and competitiveness issues that relate to how the world is mitigating and adapting to climate change. Business engagement is critical for developing efficient, effective solutions to the climate problem, and we believe that companies taking early action on climate strategies and policies will gain sustained competitive advantage over their peers. Many different sectors are represented, from high technology to diversified manufacturing; from oil and gas to transportation; from utilities to chemicals.

BELC members believe that businesses can and should incorporate responses to climate change into their core corporate strategies by taking concrete steps to establish and meet greenhouse gas (GHG) emission reduction targets. Companies are investing in energy efficiency and low and zero-GHG products, practices, and technologies. A DuPont plant in Memphis, Tennessee, uses landfill gas as a replacement for natural gas that fuels boilers and other plant equipment, replacing more than 90% of the natural gas used by the site's boilers, reducing area GHG emissions by an equivalent of the removal of 70,000 cars from the road or planting 95,000 acres of forest. Delta Airlines' in-flight recycling program successfully recycled approximately 1,108,000 lbs of material in 2010 and, through an aircraft carpet recycling partnership, recycled approximately 147,500 lbs of carpet. The tech company HP is using its core strengths in information technology to reduce the environmental impacts of dozens of industries. Leveraging IT to improve transportation

> We need to generate more long-term thinking, which would underpin our ability to encourage decision makers – such as individuals, companies, and policy makers – to use resources and talent in ways that are sustainable rather than tied to short-term gains.

efficiency at, for example, companies like UPS is significantly reducing air emissions. Using HP's video-conferencing system, HP and its customers have saved an estimated 66,000 metric tons of GHG emissions in 2 years, and HP reduced its own employee business travel by more than 43%.

The Center for Climate and Energy Solutions (C2ES) also supports companies' engagement with their employees to make a difference through the Make an Impact program, which helps businesses more actively engage their employees and communities to save energy, reduce their carbon footprint, and live sustainably. Three corporate partners, Alcoa, Entergy, and Bank of America, have reached more than 100,000 employees to encourage smarter environmental choices and empower individuals to save energy and reduce their carbon footprint. The 13,000 users of the program's "carbon calculator," which helps employees to assess and reduce the greenhouse gas emissions associated with their daily activities, have to date saved $1,300 in annual energy savings and identified 8,000 lbs in annual carbon savings, per user. While Alcoa has focused its program on educating employees to empower individual change at home, employees at the company's Intalco facility in Washington state took the message to the workplace as well. Through an employee-driven "suggestion box" process, the facility's employees were able to implement a number of efficiency recommendations that resulted in over $125,000 in energy savings in 1 year.

Companies are demonstrating their power to make a significant difference in their communities, through mitigating the impacts of their own operations, to influencing their customers' and suppliers' decision making, to engaging and educating their employees on sustainable lifestyles. This holistic approach to seeing the total set of opportunities associated with a company's business practices could be enhanced with a longer-term perspective on not just environmental and sustainability challenges but on the nature of business growth and profit.

Some companies and governments are already taking some steps in that direction, to value long-term sustainable prosperity over short-term gains. A number of major companies, including Berkshire Hathaway, Citigroup, Ford Motor, and Google, have at times chosen not to issue frequent earnings guidance to investors, a brave effort to taking a longer view in a Wall Street culture that puts a premium on short-term gains. Some countries, including Denmark, South Africa, and the United Kingdom, are now requiring that companies listed on their home stock exchanges file annual financial reports that also include sustainability risks and opportunities. Due to years of investor engagement with the U.S. Securities and Exchange Commission to encourage corporate disclosure of material sustainability risks in corporate financial filings, in January 2010, the SEC released guidance on climate risk disclosure that will promote greater corporate transparency on this issue. If "[s]ustainability is a journey, not a destination," as Charles Holliday said when he was chairman and CEO of DuPont (now chairman of Bank of America), then environmental, social, and economic challenges require a longer-term perspective and an integrated, holistic approach.

***

Meg Crawford *is Markets and Business Strategy Fellow at the Center for Climate and Energy Solutions, working with the Business Environmental Leadership Council – a group of (largely Fortune 500) corporations that partner to address issues related to climate change. Earlier, Crawford was a manager of Corporate Accountability at Ceres, a national network of investors, environmental organizations, and public interest groups working with companies and investors to address sustainability challenges such as global climate change. She holds a M.S. in foreign service and a B.S. in foreign languages from Georgetown University.*

Chapter 26
Put It on Paper: Lowering Healthcare Costs

Una Ryan

I was born in an air raid shelter under Bangsar Hospital, Kuala Lumpur. My mother and I scrambled to the docks in Singapore to try to get aboard a ship leaving for England. The trouble was that, in the "fog of war," nobody had time for records — or recordkeeping. I didn't have a birth certificate, passport, or any official documents, and neither I nor my Chinese mother looked English. My English father was captured, and I did not see him again until I was nearly 5 years old, after he had endured all that time as a prisoner of war. At the docks in Singapore, my father had told my mother to let "them" know we were English and she wisely told him to "put it on paper" because she wasn't sure that anybody would believe her without her husband at her side. After 18 months in refugee camps across the world, I finally enjoyed a happy childhood and good education in England, where I was almost never required to show my documents again. Looking back on it, I recognize how valuable a piece of paper can be, if it has the right information on it.

Paper is an everyday item of much utility, but very few treasure it: Who collects the blank paper on which they tried and failed to write a worthy poem or essay? Like many things, paper is really only as good as what you do with it. Many of the great lasting concepts of civilization were committed to paper and will endure through preservation, reproduction, or memory. Other uses of paper have huge personal significance, such as a document that can get one across a border or that proves some identity or affiliation. There are also some uses of paper that are ephemeral: paper kites, tickets, toilet paper, and more. So could this inexpensive disposable material provide the blueprint for a better, cheaper

> Information is valuable if it is timely and can be used to generate something sustainable, something that survives beyond its creation and its creators to be such a force for good that nobody can remember why it was needed in the first place.

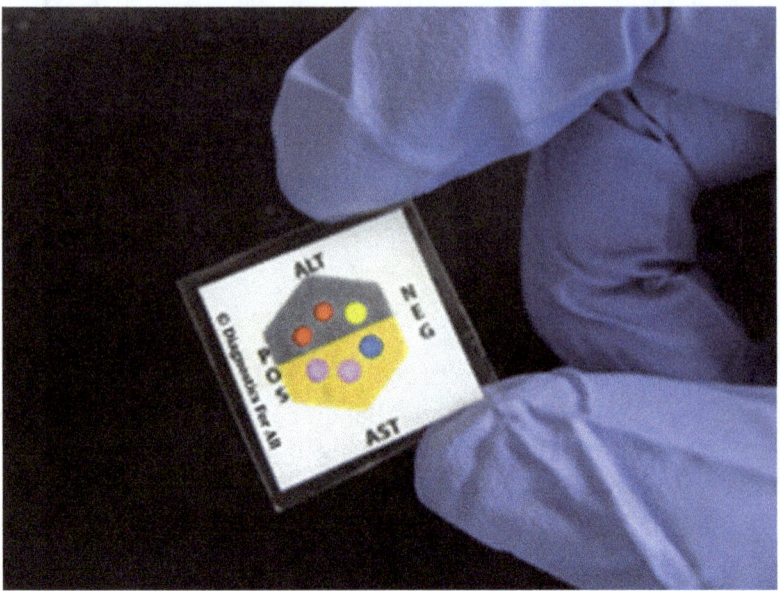

Fig. 26.1 The DFA diagnostic device is the size of a postage stamp and works by wicking a body fluid such as blood, urine, sweat, saliva, or tears through channels printed on the paper. A color change occurs when the fluid reaches the predesignated spots. The devices are called TOPS; the one shown indicates liver damage that may result from certain drugs used for the treatment of HIV or TB

healthcare system? Could printed paper be used to unlock the information needed to underlie a better understanding of global health and disease?

Imagine my excitement when I realized that the technology could be at our fingertips — literally as easily as one can prick a finger! At Diagnostics For All (DFA), a nonprofit organization, we develop postage-stamp–sized, point-of-care diagnostics designed specifically for the developing world. We call them TOPS — "test on paper substrate." TOPS can be designed to indicate many conditions or infections. A particular test will change color when exposed to a single drop of blood from a finger-stick or other bodily fluid, such as urine, sweat, stool, saliva, or tears.

DFA's tests are made of paper and based on the principles of microfluidics (see Fig. 26.1). The technology was first invented at Harvard University (see Martinez et al.: Analytical Chemistry 82, 3–10, 2010). This work has been further enhanced with inventions from DFA scientists The secret lies in the fact that paper has the capability to wick fluids without the need for external power. The paper is patterned with printed channels, so that samples applied to TOPS are wicked through the channels by capillary action; TOPS are preloaded with reagents for specific tests and use while-you-wait colorimetric readouts that can be compared to a reference guide. Using multiple layers of paper and multiple channels, the technology can perform several tests on a single chip with a turnaround time of minutes.

Diagnostic devices are not widely used in the developing world because they are regarded as too expensive or not reliable. In order to be adopted in the developing world, new diagnostic approaches must satisfy some very tough requirements. Devices will need to be accessible and affordable to people in low-resource circumstances, whether rural or urban, where reliable electricity, access to clean water, and trained personnel are nonexistent or in short supply. TOPS are expected to cost only a few cents because paper and printing methods are both ubiquitous and cheap. In this way, little bits of paper can perhaps provide a simple-to-use solution to managing the health and treatment of people in rural villages who live beyond the reach of modern medical centers or centralized diagnostic laboratories. Similarly, there are growing populations in big city slums and urban areas who require access to affordable healthcare and are frequently not served by facilities available to the wealthy.

> Sustainability is appealing because it is bigger than self and enriches the people for whom it was created, not the people who invented it.

Innovative technology such as that provided by printed paper chips can be made at a very low cost-of-goods; however, we will have to tackle other problems to achieve sustainable impact. The final diagnostic devices will need to be manufactured at large scale and be affordable in very low-resource situations, easy-to-use, portable, and disposable as well as sensitive, specific, and reliable. This is a long list, but a list that we hold out great hope that TOPS can challenge. However, in order to manage the health not only of individuals but also of populations and to begin to tackle the enormous problems of improving access to healthcare worldwide, we will need accurate intelligence about individual and population health trends as well as the involvement and empowerment of the patients themselves.

Once again perhaps we can count on the ability of paper printed with the right channels and patterned with the right information to answer both challenges. We first have to gather accurate and representative information on the diseases, people, and populations at stake. This means we have to know what we are treating. A lack of diagnosis in the developing world not only leads to a lack of treatment but also to treatment with inappropriate drugs and unnecessary side effects. Since accurate intelligence is the first step rather than the final solution (which must be treatments or prevention), the diagnostic step should add negligible cost because healthcare resources will have to stretch beyond the treatment period – into monitoring and surveillance of health and disease on an ongoing basis.

All of DFA's technology is telemedicine-compatible. The easy color readout can be compared to a printed color reference scale on the packaging; combined with the almost ubiquitous use of cell phones with camera capability, this means that accurate patient records and worldwide disease tracking become available and easily affordable. Some would call this postmarketing surveillance; others would see it as part of a global exercise in forecasting and tracking pandemics, epidemics, and endemic reservoirs of disease. Yet others would see this as lifetime wellness monitoring on a population level.

The second trend that should have a powerful effect on the sustainability of healthcare access and cost will be the move to point-of-care testing. This will lead to empowerment of patients in terms of management of their own health and treatment options. Governments cannot be expected to provide their healthcare ministries with tools that are not cost-effective. Here again, even where clinical infrastructure is lacking in the developing world, the almost universal use of cell phones coupled with the ease of use of TOPS point-of-care testing may allow tracking of health and disease on a massive scale.

> The adjectives "simple," "inexpensive," and "useful" describe the tools that people need. People will take care of the rest.

Diagnostics For All is dedicated to delivering our technology affordably to all who could benefit from it, and we have prioritized health needs in the developing world. We have begun to expand beyond human health to supporting livestock and agricultural uses of the technology in order to help small-holder farmers manage their herds and crops. However, "For All" can include benefitting those with some discretionary income. The low-cost, user-friendly, camera-ready, at-home or point-of-use tests mean that DFA technology can benefit efforts to reduce healthcare costs and improve livelihood worldwide. TOPS therefore have a number of attractive applications for developed world uses, including human health, animal health, food safety, environmental, military, and homeland security uses. In order to capture the value for developed world applications as well as to provide sustainability for our mission of ensuring affordable diagnostics in resource-poor areas, DFA has developed a business model with both nonprofit and for-profit arms. In the developing world, DFA is interested in partnering with local companies, government health systems, or locally effective NGOs. In this way, we hope to create jobs and to gain expertise from local partners into preferred distribution channels and to understand regulatory requirements on a country-by-country basis. In the developed world, Paper Diagnostics (the for-profit subsidiary of DFA) can engage in license arrangements with commercial entities that can provide knowledge of commercial markets, manufacturing, and regulatory expertise in markets beyond DFA's current scope and range.

At the end of the day, our hope is that DFA will provide three lasting benefits:

- Demonstration that a correct diagnosis leads to a life saved, over and over again.
- Blueprint for a sustainable business plan that can lower healthcare costs: point-of-care diagnostics with the benefits of high-tech but at low-tech prices.
- Recognition that the common, ordinary things of life like paper, with the right information imprinted upon it, can become the passport to a better life.

Then my makeshift birth certificate, written on that scrap of paper and sewn into the hem of my mother's coat, will have truly earned its keep.

***

Una Ryan *is the CEO of Diagnostics For All. She has an extensive background in leading biotech companies and was formerly the president and CEO of AVANT Immunotherapeutics, Inc., a publicly traded biopharmaceutical company developing vaccines. Ryan holds a Ph.D. in cellular and molecular biology from Cambridge University and B.S. degrees in zoology, microbiology, and chemistry from Bristol University, where she also received an honorary Doctor of Science degree. She was a Howard Hughes Medical Institute Investigator and has received the Albert Einstein Award for outstanding achievement in the life sciences. For her services to the research, development, and promotion of biotechnology, Her Majesty Queen Elizabeth II awarded her the Order of the British Empire.*

Chapter 27
Mind the Gap: A Different Take on Sustainability

Matthew Taylor

It has been said that the really significant divide in politics is not between the left and right but between optimists and pessimists. At the RSA–the Royal Society for the Encouragement of Arts, Manufactures, and Commerce –I chair many events at which public intellectuals give their various prophesies. On issues of sustainability there certainly is a divide between those who think that technology and human ingenuity will solve tomorrow's problems just as they did yesterday's and concerned environmentalists whose conclusions tends to be some version of "We can't go on like this."

Perhaps it is just that I am a pragmatic middle-of-the-road kind of person, but I find neither account satisfactory. Of the optimists, I want to ask, "Why should we assume this problem can be solved, perhaps it is different?" or "Haven't you read your Jared Diamond? Civilizations do collapse and precisely because they failed to address a crisis which appears in retrospect to have been staring them in the face," or "Yes, we may solve our problems in the end but at what cost in human suffering and waste?"

To the pessimists, I want to ask, "Why is it that we have been saying for so long that things like oil were going to run out but then there always seems to be more?" or "How do you reconcile your humanism with such apparent pessimism of human beings to find solutions?" and–although this I say *sotto voce*–"Aren't you concerned that you sound almost pleased at all the bad news you have to share?"

> Sustainability means each of us having a better understanding of our behavioral dispositions and marshaling that knowledge in order to lead better, more fulfilling lives.

This may be why I developed a more modest view of the future. I suggest that the United Kingdom faces what I inelegantly call a *social aspiration gap*. The problem for this country is not that people generally have radically different ideas of the kind of future they would like for themselves and their society. We all want to live somewhere that is economically comfortable, that protects and expands freedom but also avoids gross inequality, with decent public services and a peaceful and tolerant public sphere, and, yes, we do want to safeguard our environment and play our role in saving the planet. The problem is not that we disagree about our aspirations; it is that we are unlikely to fulfill them if we don't significantly change some of the ways we think and act; that is the social aspiration gap.

Let me offer some concrete examples: Successful countries balance short-term demands with long-term investment. Leaders have to be able to make difficult decisions, creating losers as well as winners. Arguably, right now enlightened autocracies like China are better able to do this than mature democracies. The democratic conversation needs to be more substantive, honest, and two-way.

The U.K. health service is facing its tightest budget settlement for many decades and in the longer term faces a growing burden in the form of chronic conditions affecting an aging population. Yet one of the biggest drivers of demand for healthcare is our own lifestyle, drinking too much, eating too much, exercising too little, not managing long-term conditions.

For Britain to be a successful economy, we need citizens who are well educated, creative, and risk-taking. There are things government can and should do to shape tomorrow's citizens, but this also involves a shift in our national culture so that we prize invention over mere accumulation and see learning as a lifelong habit.

The trade-off point between economic growth and environmental sustainability can be much higher if we voluntarily take actions to reduce our carbon emissions and unnecessary waste. Closing the social aspiration gap is an important part of any strategy for environmental sustainability.

Before I go on, I need to address an often tacit but powerful objection to this very idea that we–the people–can choose to change the way we are. Over the last few decades, thinkers on the sociological left and economic right have shared a common prejudice: Public ideas don't really matter. On the left, the ideas are seen as merely an epiphenomenon of deeper social forces; on the right, culture is marginalized in a theory that sees the efficient society as one comprised by individual preferences revealed in market transactions. Of course, ideas are more powerful if they intersect with social forces, and naturally they are more likely to succeed if they coincide with our personal predispositions. But this isn't the end of the story.

Let me offer two examples of significant social trends that are inexplicable without recognizing the way ideas can change the world. The first is

> As someone who leads an organization dedicated to achieving social progress and human fulfillment, sustainability and a consideration for past, present, and future generations have to be bound into every decision.

the transformation in attitudes to homosexuality. It is hard to explain this in terms of shifts either in individual preferences—why would so many more people suddenly become gay or why would so many people now be tolerant when their parents were viscerally prejudiced? —or in underlying social forces—capitalism seems just as able to thrive in homophobic as in liberated societies. Instead, an important part of the shift comes from the way the gay community and its supporters responded to the threat of AIDS; rather than retreating into the shadows, the response was one of mobilization, pride, and self-help.

A different example is the growth of Fair Trade. Did we suddenly grow a social conscience? Did capitalism need a new market in ethical goods? Or was it that crusading leadership, business skills, and social organization found a way to tap into human altruism so that an idea that had for many years been largely confined to stalls outside churches moved to the aisles of every major supermarket?

One reason ideas matter for society is that human beings and human behavior are complex. Over recent decades, a variety of disciplines, ranging from neuroscience to evolutionary psychology to behavioral economics, have undermined the reductionist view of human beings as mere agents of impersonal forces or of society as being nothing more than the aggregation of possessive individualism.

We know that most of our actions are instinctive and automatic, not the consequence of rational and conscious calculation. Indeed, we know that often conscious thought is a confirmation of an automatic impulse rather than what drives our behavior. We know also that we are profoundly social beings. Role-playing experiments show how much our personality and our judgments can change if we are placed in different contexts with different prompts and norms. Long-term social network analysis shows our behaviors and attitudes can be affected by changes among those in our circles of networks even at three degrees of separation.

We also know that we are in many ways idiosyncratic. We find it hard to bring the long term into the present and to do the things we ought to. Directly contradicting conventional economic theory, it appears that monetary incentives actually inhibit our performance in complex tasks. And we are misguided; in many circumstances human beings are not very good at assessing their own abilities, at predicting their future, knowing what will make them happy, and even accurately recalling what made them happy in the past. On the positive side, we also know that most human beings seem to have an innate capacity for empathy and sense of fairness.

Reflecting both our eighteenth-century origins and our modern mission, the RSA's new strapline is "21st-century enlightenment." This idea combines our analysis of the social aspiration gap with new thinking about human nature and motivation and adds a philosophical dimension.

The original Western enlightenment was a complex and contested process, but at its heart lay three revolutionary ideas: autonomy (the idea of human freedom), universalism (the idea of that all human beings are deserving of dignity and rights), and humanism (the idea that society should be organized not according to the rules of kings or bishops but to maximize human fulfillment).

The "21st-century enlightenment" approach suggests we need to reexamine and refresh the way we think about these ideas. On autonomy, we need to replace the idea that freedom is delivered through possessive individualism and instead promote the idea (one long propagated by the world's religions) that genuine autonomy comes through deeper self-awareness and self-discipline. In relation to universalism, we need to be slightly less focused on the content of universalism—which rights, entitlements, capabilities?—and more on the foundations of universalism; what is it that fosters solidarity and the capacity for empathy we need in a shrinking world? And in relation to humanism, we need to see past the compelling modern logics of markets, bureaucracy, and science and technology to provide the spaces for a deeper, more ethical and humanistic debate about the kind of future we want to build in our neighbourhoods, countries, and world.

> Relying on tired old binaries of "good" versus "bad" will always sound appealing in the debates concerning sustainability, but unfortunately that will not get us very far.

I began with pessimism and optimism. Over recent decades a powerful trend in many Western nations has been the steady rise in social pessimism: While people tend to be unrealistically optimistic about their own and their family's prospects, they tend to be too pessimistic about the ability of society to change and advance. For example, not only did few people in the U.K. or U.S. predict the decline in crime and certain other social pathologies that have taken place over the last decade, but many still refuse to believe them.

Social pessimism is an enemy of progressive thought. To believe we can create a sustainable future, and not just rely on technology to come to our rescue, requires us to believe we can choose to change. Just as the first Western enlightenment was ushered in by scientific breakthroughs that challenged religious doctrine, so new thinking about human nature can help challenge deterministic or individualistic accounts of human nature.

The question is how, then, do we turn that new understanding of human nature into an action plan for sustainability? Part of the answer will always lie in our institutions. The families we are a part of, the organizations we work within, and, of course, the state we pay our taxes to and receive our services from will always be places in which we can nurture better social norms. But I want to argue that institutions of private enterprise also have a central role to play in delivering a sustainable future.

In one of my annual lectures at RSA, I drew upon the theme of 21st-century enlightenment to argue that business can use its relationship with customers and wider society to help us live better and more meaningful lives. Indeed, the pursuit of profit can and should be combined with the attainment of social goals. For too long we have been stuck in a recurring debate about the merits of business and its role in society, with the antagonists of business seemingly always winning out.

New insights into human nature and a better understanding of our behavioral dispositions can help enliven this debate and open up new terrain for how we think about business responsibility. While corporate leaders have tended to peddle the view of humans as rational, wholly self-interested beings, marketing departments have done the opposite, preying on our cognitive frailties and shaping our preferences to suit their own ends. It is clear that business doesn't simply respond to desire and demand; they actively shape it.

While it is all too easy at this point to retreat into a critical stance toward business and particularly its impact on sustainability, I argue that we should consider people's increasingly thick relationships with companies and their brands as an opportunity, not a threat. Nike, for instance, has over the last few years wedded its brand image to fitness, encouraging its customers to have more active lifestyles and to sign up to their Nike Plus running community. Or take the home improvement chain B&Q, which has established inexpensive DIY classes across many of its stores in a bid to encourage its customers to buy better-quality materials and to mend and repair them once they fall into disrepair.

Enterprises work in challenging environments and against tough competition. Our task should not just be to deride companies, but rather to appreciate the way that business works and to engage them in a conversation about how achieving sustainability can actually go in tandem with the pursuit of a stronger, longer-lasting business.

And this cuts to the heart of the wider debate about how we can realistically achieve a sustainable future. If we truly want to be a sustainable society, we have to acknowledge that people can change their behavior for the better and to be willing to draw upon new sources, such as private enterprise, that can help aid that transition. Relying on tired old binaries of "good" versus "bad" will always sound appealing, but unfortunately will not get us very far.

***

Matthew Taylor *is chief executive of the RSA in London, U.K. Prior to that, he was chief adviser on political strategy to the British Prime Minister. Earlier, he was the director of the Institute for Public Policy Research, Britain's leading center-left think tank. Taylor is a frequent media commentator on policy and political issues and has written for publications including the* Guardian, Observer, New Statesman, *and* Prospect. *He is a regular panelist on BBC Radio 4's* Moral Maze.

Chapter 28
Sustainable Scientific Research

Katepalli Sreenivasan

Sustainability has a variety of different connotations, but is often related to global resources – sometimes constrained by a country and the context. My focus here is the sustainability of doing *high-level science* in a developing country. This facet of sustainability is less well discussed because it depends largely on a nation's long-term vision and goals for its citizens, the external image it wishes to project for itself, and the resources it can muster in support of its aspirations. But it also depends on the collective will of scientists as an international community. In the final analysis, this will be my main point.

For a reason that is explained later, I have had the opportunity to visit many scientific laboratories in developing countries. Each laboratory is special and its necessities and problems are different, but a theme frequently witnessed is that many pieces of equipment would be seen to be either in poor working order or in unopened boxes. The reason is often that the expertise needed to operate the equipment and keep it functional did not exist. It was no doubt in part due to the paucity of money but quite often due to a lack of willingness to learn: The technical knowledge needed to maintain advanced equipment did not exist at a sustainable level. There might have been one person who knew how to operate it at one time, but she would long ago have migrated to greener pastures; there might have been an active scientific collaboration with another country at one time, during which the equipment was indeed working well – but its use waned when the collaboration ceased – and so forth. I cite this situation merely as the tip of a large iceberg,

> Sustainability has a variety of different connotations, but is are often related to global resources – sometimes constrained by a country and the context.

one that reflects in many countries the deeper problem of being unable to develop skills for nurturing homegrown researchers and research institutions.

The homegrown ability to do advanced science demands that the country in question must invest adequately in science education at all levels, including the Primary and the secondary; equally important is the training of an adequate cadre of inspiring teachers. These commitments require the support of the citizenry as a whole, which ought to understand the role of science not merely as a precursor to technological advancement but as a cultural necessity, without which one cannot be a self-confident citizen of the twenty-first century.

Unfortunately, countries in which education and research have not been thriving side by side for many years cannot expect to develop education and research frontiers linearly one after the other: Primary education today, secondary education 4 years from now, and so forth, eventually leading to high-level science some 20 years on. For one thing, the world will have moved in 20 years beyond anything we can imagine. For another, the linear development strategy does not work even if there were enough time: Planning well for a lower level often requires the presence of some excellence at a higher level. And lower levels of education, whose importance can never be exaggerated, are arduous to rejuvenate because of the enormous numbers involved. It is thus clear that one cannot wait for lower levels of education to be "fixed" before a plausible research enterprise can be conceived.

Indeed, we already have ample evidence that the exclusive focus on lower levels of education while neglecting higher education has led to catastrophes in the 1980 and 1990s. The African experience (if one may generalize momentarily this essentially nongeneralizable scene) is that when those countries became independent in the 1950 and 1960s, they inherited the colonial system, whose main task earlier had been to provide a sustained supply of civil service personnel for the government. Several countries underwent political turmoil and various forms of civil wars in the 1970 and 1980s. At that point, the World Bank began to take great interest in education in African schools and universities and argued that universities are privileged enclaves whose returns do not warrant the costly investment and support. Faced with this advice, on the one hand, and other immediate priorities of national governments, on the other, the support for

> The challenge of sustaining a high-quality research laboratory in many countries reflects the deeper problem of not being able to develop skills for nurturing homegrown researchers and research institutions.

higher education was decimated and the focus turned toward lower levels of education. The World Bank lost track of the fact that universities play an important role as the vehicle of development of a country and in defining the ideas and ideals that make up the fabric of a society, including the definition of its lower level of education: Just imagine the U.S. without Harvard and Stanford, or the U.K. without Cambridge and the Imperial.

A quote may be useful in amplifying my remarks. The paragraph below is part of a statement by Ann Therese Ndong-Jatta, Secretary of State for Education of the Republic of the Gambia:

> A condition for qualifying for World Bank assistance in the education sector was for African countries to divert resources from higher education and channel them instead towards Primary and basic education... Needless to say, with the tremendous pressures that come along with World Bank and IMF conditionalities ... higher education in Africa virtually went under. To this day, many countries have not been able to recover from that onslaught on African higher education. Some of our finest institutions have thus almost been destroyed, thanks to the imposition of bad policies from partners....

The need to work on all fronts together is obvious, but it is equally obvious that the situation in many countries is dire enough that building research infrastructure and attaining international standards (whatever exactly it may mean) cannot be attempted easily: One problem is often the financial diversion that this creates from what is regarded by politicians as the bread-and-butter issues; the other problem is that human resources needed to carry out scientific research simply do not exist, or cannot be created in a sustained manner. The financial part of this dual need is mainly the country's responsibility, with help from international organizations such as the World Bank, IMF, the Forum of G8 countries, and so forth. This part is dictated by the priority that a country sees for its future and cannot be imposed or conceived from outside. But what one can do from outside is to raise the awareness of the politicians and the populace and help create a proper environment where the importance of scientific progress is understood.

On this second aspect of creating scientific capacity, it is obvious that the international scientific community has an important role to play. But why should the scientists be interested in this task? How should they be engaged? These are the questions I shall address very briefly.

The international scientific community might sometimes be persuaded by the ideal of altruism and the argument that a normal part of being a scientist is to take some responsibility toward developing countries in all corners of the world. After all, science is "universal" and scientific talent is presumably distributed uniformly across populations. But this argument carries only a finite weight because the average scientist is very busy in arranging his own scientific career to prosper. A more compelling case rests on the following two grounds:

1. There is a rising need to engage an ever-expanding cadre of scientific communities to make significant scientific progress anywhere. No single country has all the brain power and the financial prowess needed to make all scientific progress to happen. The clear examples are Big Science experiments such as the Large Hadron Collider at CERN and the International Thermonuclear Experimental Reactor in Cadarache, but they are not the only ones. Indeed, trends are that some developing countries such as Brazil, China, and India, and others such as Singapore and South Korea, are expanding their science base much more rapidly than the advanced nations of today, opening up new scientific opportunities for all.

2. The increasing interlinking of different parts of the world ensures a common interest in solving numerous and commonly shared problems associated with climate change, water scarcity, degrading ecosystem, spread of infectious diseases, international terrorism, migration of large masses of po pulation, lack of food security, deep economic recessions caused by connected economies, and so forth. Today, the prosperity of one part of the globe depends far more acutely on the survival and well-being of all other parts of the globe as the place for decent human habitat. While inherited riches are less easily distributed, sharing knowledge – to which many civilizations have contributed, and which is at the root of being able to build self-sufficiency – should not be obstructed. (Political masters take exception to this basic argument, and one should discuss each of those exceptions rather than denounce the general rule just stated.) A compelling case can be made that everyone thrives by sharing knowledge rather than by selectively preventing access to it.

International scientific organizations have made considerable progress in providing access to modern scientific knowledge. My own former institution, the Abdus Salam International Centre for Theoretical Physics in Trieste (ICTP), Italy, continues to play an important role. Despite the implied narrowness of its name, ICTP regards all of science to be within its mandate and is learning to navigate new challenges as they arise. The main challenge

> One problem with the establishment of a research infrastructure in poor countries is the financial diversion that this creates from what is often regarded by their politicians as the bread-and-butter problems; the other problem is that the human resources needed simply do not exist or cannot be created in a sustained manner.

is to engage larger and stronger science communities in scientific capacity building and to convince local governments that this enhanced scientific capacity can be harnessed for economic development.

In the task of building scientific capacity, a very significant role is played by regional scientific networks, which support each other in ways that no single institution by itself can accomplish. These networks work best among countries with political or cultural ties, or in geographic proximity. For example, it makes greater sense for the U.S. to take active interest in science in South and Central America and for Europe to take a similarly greater interest in Africa. Needless to say, such an interest must be benign, not overbearing.

In the final analysis, even in these days of overwhelming international connectivity, each country must solve its own problems – including investment in science and technology. This situation is often governed by the necessities of sovereign territories. On the other hand, creating scientific capacity in any country is the responsibility of all countries, especially of those with natural connections of past political ties or present geographic nearness or likeminded cultural affinities. It is only through a competent scientific capacity that a sovereign population can apply itself to solve its problems, thus contributing to the well-being of this world as a whole. Sustained scientific progress can contribute to true sustainability.

***

Katepalli Sreenivasan *is senior vice provost and university professor of physics and mathematical sciences at New York University (NYU), and the provost of Polytechnic Institute of NYU. He served as the director of the Abdus Salam International Centre for Theoretical Physics in Trieste, Italy, for a little less than 7 years. Sreenivasan has been elected, among other honorific societies, to the U.S. National Academy of Sciences, the U.S. National Academy of Engineering, the American Academy of Arts and Sciences, the Indian Academy of Sciences, the Indian National Science Academy, the Academy of Sciences for the Developing World, and the African Academy of Sciences. Sreenivasan's research expertise is fluid dynamics in a broad sense and has touched a few other areas of applied physics. He is recipient of a Guggenheim Fellowship, the UNESCO Medal for Promoting International Scientific Cooperation and World Peace, as well as several honorary doctorates.*

Chapter 29
Energizing Sustainable Development

V.S. Ramamurthy and Narendar Pani

In the foothills of the Himalayas, there is a region with a large number of springs, some of which are perennial. For hundreds of years, people have been using the running water to turn small mills used for dehusking and grinding wheat. One organization, Himalayan Environmental Studies and Conservation Organization (HESCO), realized that there is not only a significant scope for efficiency improvement of the mill itself, but that restricting its use to grinding wheat alone was actually limiting its use for a few hours in a day while the stream was flowing all the time. HESCO carried out a technology up-gradation exercise wherein ball bearings were introduced, the wooden paddles were replaced with metal paddles, and the mills were coupled to a generator so that when they are not used to grind wheat, they produced electricity for the village. For the first time, a village far away from the main national electricity network had access to electricity to light their homes, to power small machines, and a host of other uses. The cost for a 10-kW system was an unbelievable Rs. 50,000 (approximately US $1,000). Over a period of time, HESCO was able to bring into their fold several thousand watermills for up-gradation.

The emergence of the springs as a source of electrical energy revived interest in the springs themselves. The springs in this region are generally seasonal and are derived from seepage waters flowing through the shallow weathered and fractured zones. In collaboration with the scientists of Bhabha Atomic Research Centre,

> Sustainable development is development that does not take more from the environment than nature can replenish.

Mumbai, HESCO was able to track the subsoil transport of water using isotope hydrology techniques:

> Based on local geology, geo morphology, hydrochemistry and isotope information, the possible recharge areas were inferred. Water conservation and recharge structures such as subsurface dykes, check bunds and contour trenches were constructed at the identified recharge areas for controlling the subsurface flow, rain water harvesting and ground water augmentation respectively. As a result, during and after the following monsoon, the discharge rates of the springs not only increased significantly but also did not dry up during the dry period. (Shivanna et al.: Current Science 94(8), 2008).

The ecological sustainability of this experience is immediately apparent. The entire exercise does not take away more from the environment than nature can replenish. It can remain ecologically sustainable even as it expands to cover many more villages, as long as what it takes away from the environment can be naturally replenished. But man does not believe he lives by nature alone. The natural aspiration of every human being, whether she is from the developed world or from the developing one, is to have a better quality of life tomorrow than she has today. Unfortunately, there is no universal definition of quality of life, with the result that even the Queen of England can feel poorer than Bill Gates and aspire to improve her lot. These aspirations are not confined to what may be termed the fundamental, non-negotiable needs of every human being – food, water, and habitat. There are also needs associated with lifestyles. The varied and often opulent lifestyles thrown up by social, cultural, and economic changes are not all benign. The realization of all these depends on the availability of some natural resource or the other. The unbridled growth of individual aspirations, and their realization through development, inevitably draws out of the environment more than nature can replenish. As Mahatma Gandhi is widely quoted as saying, "The earth provides enough to satisfy every man's need, not every man's greed."

> The only development option available to countries that are not at the top of the economic hierarchy today is one based on sustainable consumption of natural resources, sustainable management of waste (nature does not produce any waste), and sustainable strategies for the resolution of social, cultural, and economic conflicts.

And there are now signs that the greed has gotten to a point where the earth can no longer sustain it. We are already seeing signs of nonsustainability in several areas. The acute food shortages in some part of the world and the increasing concerns on the long-term availability of energy resources and fresh water are all signs of this nonsustainability. There is some hope that technology will help us out of this nonsustainability by raising the efficiency of our use of natural resources. The same land and water can be made to give us more food; the same energy to give us more output, and so on. But while technology does offer new avenues to combat this nonsustainability, it also brings with it new concerns, including the environmental concerns associated with increasing demand for energy. It is then not enough to simply increase the efficiency of technology; it is also important to do so in a way that does not throw up new challenges to nature.

Even as we are sensitive in assessing new challenges to nature, though, it is important not to take too narrow a view of nature. In the intensity of debates on the environment, it is tempting to see this question entirely in terms of a man versus nature perspective. Too often there is an implicit perception that man and nature are two entirely different categories in conflict with each other. Protecting the environment then becomes a question of slowing down growth even if that slows down the material benefit to mankind. But it can also be argued that man is an intrinsic part of nature. Birth, life, and death affect nature, and mankind is no exception. The task of protecting nature then includes protecting the basic survival of the most vulnerable sections of humanity. And technology can play a critical role in this process. Fortunately, it is possible through localized technological strategies to address the material concerns of the most vulnerable in a way that does not do too much damage to the environment.

> A more reasoned and explicit evaluation of all technological options, without being trapped in one or the other dogma, would be required to improve the broader appreciation of sustainability.

No matter how economically and environmentally efficient these technological interventions are, they come up against the hurdle of perception. Such initiatives are necessarily related to specific problems in small areas, making them relatively small in scale. They do not have the grandeur in the development picture that a large project has. But as the world gets more sensitive to the challenge of sustainability, it becomes more difficult to ignore the fact that the large project is taking out far more from the environment than nature can put back. The little peaceful villages in the foothills of the Himalayas doing all the right things by the environment, with their technologies drawing out far less from the environment than nature can replenish, may seem a quaint example. But behind the quaintness and the sense of helping the vulnerable, these examples are a reminder that small drops do finally make an ocean. And an ocean of development that brings with it the value of helping the vulnerable even as it retains the pristine beauty of nature makes sustainability a cause well worth striving for.

***

V.S. Ramamurthy, *a nuclear scientist, is professor and director of the National Institute of Advanced Studies in Bangalore, India. He has served as a senior administrator at the Bhaba Atomic Research Center, Mumbai, and as secretary to the Government of India in the Department of Science and Technology. In these roles, he has been involved in the promotion of scientific research and development in India for over two decades. He has received Padma Bhushan, the third-highest civilian award from the President of India.*

Narendar Pani, *an economist, is a professor at the National Institute of Advanced Studies and an adjunct professor at the Indian Institute of Science in Bangalore, India. His books include* Inclusive Economics: Gandhian Method and Contemporary Policy.

Chapter 30
The Importance of Sustainability in Helping the Poor

Mechai Viravaidya

There are many perceptions of how sustainability should be applied to businesses, governments, militaries, families, and other institutions. However, the nonprofit, nongovernmental sector that endeavors to help solve some of the century's most pressing problems has not taken a critical look at long-term financial sustainability. Many nonprofits are dependent upon an ad infinitum generosity of wealthy donors to continue their operations. The very nature of helping the poor is not financially sustainable, and during the recent financial crisis, philanthropic support of the nonprofit sector has declined. To ensure the long-term sustainability of organizations that work with the poor, maintaining a business arm to provide financial support is instrumental toward the continuity of operations.

> Sustainable development is the ability to carry out the important business of helping the poor by encouraging independency – for both the beneficiaries and the implementing organization.

Consider the case of a generous donor who has offered $1 million toward a nonprofit organization that works on professional development of teachers at underprivileged schools. The organization and donor have agreed to allocate the funds on renting space for the workshops, salary cost for the trainers, materials for the workshop, logistical costs for the trainers, and other expenses. The workshops could commence immediately; however, over time, the $1 million would be depleted, and the nonprofit organization would need to come begging to the benefactor for another contribution. In the meantime, no training for teachers in underprivileged schools would be able to take place as the nonprofit organization scrambles to obtain the necessary funding. This cycle of begging and spending, and then begging some more, is typical of the nonprofit sector, whether the donor is a government body, a large foundation, or a wealthy businessperson. This paradigm needs to be fundamentally altered.

Rather, if the donor had agreed to provide the $1 million as capital investment toward a business associated with the nonprofit, then the donation would be have been much more financially sustainable in the long term. Establishing a profitable business to support the endeavors of a nonprofit, often called a social enterprise, is a much more attractive philanthropic investment, because the donor knows that his or her contribution will provide long-term sustainability. It is necessary to separate the business arm and the nonprofit arm in such an organization, but that does not mean that the core competencies cannot be shared. For instance, the business in this hypothetical case could have involved the nonprofit providing consultant work for universities and international schools, which would be profitable. The profits would then be used to help the poor at no cost, and the nonprofit would not need to worry about securing donor funding to manage its operational overhead.

> Sustainability is appealing because no one wants to be a beggar, dependent on the generosity of others.

Governments around the world have been slow to recognize these unique social businesses and develop appropriate legislation that rewards them with appropriate tax benefits. However, many innovative social entrepreneurs have utilized profitable business to support their nonprofit organization. In 1974, we began the nongovernmental organization Population and Community Development Association to address the need for family planning in Thailand for the poor. After one year of operation, we realized that we could not continuously approach our generous sponsors and ask for more money. Therefore, we applied for a loan to operate a health clinic that provided basic services to Thais. We were able to pay back the loan within a year, and the clinic continues to be in operation to this day in Bangkok.

Since our humble beginnings, experimenting with the model of having both a business arm and a nonprofit arm, we now operate 18 social enterprises, including our Cabbages & Condoms Restaurant chain and our Birds & Bees Resort in Pattaya. Rather than the profits going to benefit the shareholders, all profits from these enterprises must be used for business expansion or go toward the operational costs of our efforts to serve the poor in Southeast Asia in the area of HIV/AIDS prevention, poverty eradication, environmental awareness, democracy promotion, gender equality, and education. Through our social enterprises, we fund approximately 70% of our costs, with only 30% coming from donors. This is the model we use to ensure financial sustainability, ever during difficult economic times such as today.

We would be hypocrites if we preached a model of financial sustainability and then adopted a welfare approach to our endeavors to help get the poor out of poverty. Handouts are frequently the typical approach of governments around the world, who often are seeking to get votes around election time. Many years ago, we realized that the poor remained poor due to two things: lack of business skills and no access to credit. Walking the streets of Bangkok or the dusty roads of rural Thailand, it is readily apparent that the poor are actually businesspeople seeking to make a profit. They operate a number of businesses, such as food stalls, selling handicrafts, and driving taxis. But they find their businesses failing due to lack of training and usurious loans from unscrupulous loan sharks.

> I would like to encourage people around the world who are in the nonprofit sector to be more entrepreneurial, recognizing the fact that the poor are businesspeople seeking to make a profit, and every organization needs to pay attention to their own financial sustainability.

To improve the quality of life for the poor, we began implementing the Village Development Partnership, which has been employed in over 500 villages throughout Southeast Asia over the past 25 years. First, community empowerment occurs by establishing a gender-balanced, democratically elected Village Development Committee within the selected rural village. Then a Community Needs Assessment is generated by the villagers to provide long-term development objectives that they wish to achieve. Microcredit loans and business skills training are then made available to rural villagers so that they can start up small businesses and overcome poverty through an entrepreneurial approach, as opposed to a welfare approach. Funding from the sponsoring company or individual goes toward initially establishing the Village Development Bank in order to provide these small loans. The initial capital is not available until the entire village has taken part in a tree-planting activity, which fosters a spirit of ownership in the Village Development Bank for these communities.

By involving the community in every step of the process, we are able to avoid the problems of underprivileged communities becoming dependent on outside organizations. Ultimately, our goal is to be able to step back from helping the poor, so that they can help themselves. By approaching the challenge of development with a focus on financial sustainability, both for the implementing organization and for the beneficiaries, the rewards are guaranteed for the long term.

***

Mechai Viravaidya *founded the Population and Community Development Association (PDA) in 1974 to address the unsustainable population growth rate in Thailand. In between running PDA's activities, Mechai was appointed to such key positions as Thailand's Cabinet spokesman, the minister of the Office of the Prime Minister, and chairman of several of Thailand's largest government-owned enterprises. He was also elected to the Thailand Senate for three separate terms. Mechai has received numerous awards and honorary doctoral degrees, including the United Nations Population Award, the Bill and Melinda Gates Award for Global Health, the Prince Mahidol Award for Public Health, the Ramon Magsaysay Award for Public Service, and the Skoll Award for Social Entrepreneurship. He was also named one of* Asiaweek's *"20 Great Asians" and one of* Time *magazine's "Asian Heroes."*

Chapter 31

Why Is Waste a Dirty Word?

Melanie Walker

Trash. Garbage. Junk. Toxin. Pollutant. Emission. These are names we ascribe to metabolites and byproducts of human consumption that no longer have value to us. As a species, we collectively sequester our unwanted outputs into landfills, ooze contaminated effluent into our water supply, and release reactive gas into the atmosphere – on purpose. With the continuing and unprecedented mass migration of people to urban settings, the volume of solid waste concentrated in small areas is on track to overwhelm the environment far more quickly than most people understand.

In natural systems, however, the idea of this waste being considered "trash" is largely inapplicable since most organisms have developed physical schemes specifically to deal with byproducts from important processes. During medical school, I was fascinated by the fact that more of the body's cardiac output was targeted at the kidneys than the brain. While both organs play important roles in human metabolism and adaptation, it fascinated me to see how a waste processing organ had evolved to such prominence within the system hierarchy. Significant morbidity and mortality can result from failure to manage human waste products, but fortunately, the human body has developed comprehensive mitigation strategies.

> Sustainability is about balance. We aren't alone on this planet, and we need to do more to ensure that our system has the capacity to maintain itself over time – starting with our own wastebasket.

For example, proteins like ubiquitin detect and destroy other proteins that are malformed or useless. Compounds called "scavengers," such as reactive oxygen species, can bind to toxic materials and eliminate them.

Plant and fungal cells (in addition to some protist, animal, and bacterial ones) contain a saclike enclosed organelle called a "vacuole," which is a compartment of inorganic and organic aqueous materials that provide water storage, but also store and process waste, maintain cellular turgor, and even serve as a structural support for the cell. Similarly, animal cells contain lysosomes, which are organelles containing enzymes to break down waste and cellular debris.

If the idea of cellular waste or molecular-level metabolism feels abstract, consider social adaptations developed at the species level. Myrmecologists have described the rigid communal hierarchy actualized by leaf-cutting ants (*Atta cephalotes*) to dispose of rubbish. Elephant excrement is an important resource within the pachydermal ecosystem, just as the post-spawn carcasses of salmon are to some aquatic habitats.

While the efficient uses of waste are not only acceptable but an important part of sustainability in nature, humans are unique in both the amount of waste produced and the toxicity and permanence of many of the unwanted and expended refuse items. Over the past few years I've been able to integrate some of my medical understanding of waste into a slightly different system: human society. A number of ideas and interventions have come to my attention, but one area I found to be of great "kidney-like" importance is human wastepicking. The notion of human wastepickers is for many people nearly unthinkable. Most of us have the great luxury of functional municipal solid waste management systems in our cities, and have never seen a landfill – much less earned a living processing trash from one. Even so, it is widely believed that about 10 % of the urban poor work as wastepickers in developing world cities sorting, transporting, collecting, and recycling waste from homes, informal dumpsites, and landfills.

In addition to being helpful to the environment and a source of income and employment for the collectors, the benefits provided by informal waste collectors include:

- *Contribution to public health and sanitation.* In many rapidly urbanizing cities, wastepickers are the only hope for collection since developing world municipalities only collect between 50–80 % of the refuse generated in their cities.
- *Provision of inexpensive recycled materials to industry.* Informal wastepaper collection in Mexico meets about 75 % of the national paper industry

> "Sustainability" has become a trendy word that means different things to different people. It is an appealing concept to me because of the tremendous scientific and technologic potential to innovate. Whether it is taking examples from other life forms or from our own species, the opportunity rests on the development of actionable ideas.

need, and raw materials are purchased at less than one seventh the price it would pay for market pulp from the U.S. Of the 2.7 million tons of plastic PET bottles on U.S. shelves in 2006, four fifths went to landfills. Yet just five PET bottles (plastic soda bottles) yield enough fiber for one extra-large T-shirt, one square foot of carpet, or enough fiber fill to fill one ski jacket.
- *Reduction in municipal expenses.* Studies in Bangkok, Jakarta, Kanpur, Karachi, and Manila demonstrate that informal waste collectors save each city at least US$23 million a year in costs for waste management and raw material imports.

I try to be environmentally conscious, but like everyone else, I must make daily decisions about where to draw the line and how to do my part. A quick survey of my consumption over the past 24 hours confirms that I'm a pretty average American that generates about 4.5 lbs of solid waste per day. Not only that, my (and likely your) annual carbon footprint approaches nearly 50,000 lbs of CO_2 annually. Is that bad? Is this level of consumerism sustainable? Maybe the seminal question is: Does society need to make it easier for people like me to do the right thing by creating the right incentives or alternatives?

> We need to change the mindset of humans: waste isn't trash. It's a resource.

In the developing world, these incentives already exist (and are captured by millions of wastepickers, catadores, zaballeen, and ragpickers), even if they were created by relatively extreme poverty. By recognizing the value in what they do and integrating their efforts into more inclusive and comprehensive waste management systems, we can make strides toward true sustainability. Innovation is desperately needed in this space too – since things like landfills, incineration, and garbage relocation service only to delay consequences or spread risk. I spend a lot of time thinking about technologies that can change the way we process waste.

What I envision as the seminal first step is a sea change in the way humans perceive waste. Although it is probably too late to stem the tide of consumption, we have abundant opportunity to transition our waste streams into resources that we want and need.

- The solid waste industry currently produces more than half of America's renewable energy, more than the combined energy outputs of the solar, geothermal, hydroelectric, and wind power industries.
- Recycling and composting 82 million tons of municipal solid waste saved almost 1.3 quadrillion BTU of energy, the equivalent of 224 million barrels of oil.
- Recycling just one ton of aluminum cans conserves more than 207 million BTU, the equivalent of 36 barrels of oil, or 1,665 gal of gasoline.

While we wait for breakthroughs in science and technology, maybe we can be smarter about our waste – otherwise, we will face the consequences of our unsustainable choices. Waste shouldn't be a dirty word.

***

Melanie Walker *is the deputy director for Special Initiatives at the Bill & Melinda Gates Foundation and clinical associate professor for Neurology and Neurological Surgery at the University of Washington School of Medicine in Seattle, Washington.*

Chapter 32
How Much Is Enough? Making It Personal

Toinette Lippe

For as long as I can remember, the guiding principle of my life has been usefulness. The aim of everything I do is not to have anything left over. I do not buy or cook more than I need. I go through my closets to see what I can pass on to others, and feel guilty if I am not using whatever I own – books, sweaters, shoes, you name it. The way to achieve this is to live so that supply does not exceed demand or consumption, to be satisfied with what you have, and, whenever possible, share it with others, not holding anything back. Trust that the universe will respond to you in the same way that you respond to it.

This may be a rather revolutionary concept to many people, but I urge you to try it: When you sit down to a meal, help yourself to no more than you are sure you can eat. You can always have more, but once something is on your plate, it tends to get thrown out if you don't finish it. Why do so many of us put so much food on our plates? It can't be that we fear starvation. Not in this generation and in this country. After almost 50 years of living in New York City, I am still uncomfortable with the amount of food restaurants serve. Has it never occurred to them that we might prefer to eat less, pay less, and weigh less?

> Using what is available rather than wanting something that *isn't*. Also, finding multiple uses for things: One day I bought golden beets at the farmers' market, painted their portrait, photographed it, used the image to make notecards, and then ate the beets for lunch.

Our experiences and possessions never seem to bring us lasting happiness or completion because we always want something more and it keeps eluding us. We try to fill the vacuum that we believe to be inside us. But we didn't come into this life to make a lot of money, chalk up experiences, or amass stuff we can't take with us when we go. Not only do we hold onto what we already have, but we want to acquire as much more as we can. I view possessions as possessing me rather than vice versa. It is not the number and diversity of our possessions that is the problem, but our attachment to them. When the attachment grows thin and the filament breaks, we discover that we do not really want so much anymore. The freedom we are all seeking is freedom from the fear of losing what we believe we own.

We have been conditioned to believe that more is better. Many people console themselves with the idea that "Less is more." But I realized lately that I don't want more. I just want enough. "Less is more" contains the subtle message that if you have less, you will receive more. It is still a promise that more is better. Yet all we need is whatever is sufficient to deal with the situation we find ourselves in. So I propose that we change the adage to "Less is enough."

Most people believe that they expend just the right amount of energy for whatever they are doing. However, if you are like me, when the phone rings, not only do you try to pick it up immediately, but you grip it harder than you need to. The verb that is generally used for phones is "cradle," but that is rarely what people do with them. Whoever is at the other end is not going to give up after just a couple of rings, so you don't have to act as though the call is an emergency. When you pick up the phone, notice how tightly you are holding it. If you can relax the muscles in your hand, that release will be felt throughout your body. Enlightenment begins with relaxation. The hint is there in the word itself: "en-lighten-ment."

We tend to put more force into every movement than is necessary. Not much effort is required to turn the handle of a door and push it open. That is the way the handle and hinges are designed. Experiment with actions like these to see how little strength is necessary. It will be a revelation. Each time we relinquish the extra effort, we will be able to save that energy for doing something else. One reason we all get so tired is that we spend more energy than we need, and much of our tension comes about because we are not content to simply perform actions. We add into them layers of feeling and desire that are counterproductive, so look for your intention and see if you can recognize it as you check on the level of effort you are using. If you can catch a glimpse of your desires and let go of them, then you can devote yourself completely to whatever it is you want to do, and your action will be untrammeled.

> Sustainability, in my view, is the natural way to live.

It is important to be aware of what is going on in our minds because whatever it is, it is absorbing our energy and attention. This constant

> Don't wait for governments or NGOs to do the work. It all begins and ends with each one of us.

activity, of which most of us are completely unaware, can be exhausting and wasteful of our resources. Whatever we give our attention to grows, so we should find out what that is. It may be fear, loneliness, anxiety, or any number of things, but it behooves us to take a look. If what we had planned to do was go for a walk, why not just walk? It could be fun, but if all we are doing is continuing an inner conversation that has been running all day, the walk will not refresh us.

People tend to look for distraction – anything that will take their minds off whatever they don't want to think about. This often takes the form of "entertainment." Most people will do anything to avoid being where they are, doing whatever it is that needs to be done. And sometimes they are seduced (by themselves) into thinking that whatever anyone else is doing must be more interesting. This assumes many guises, but all of them make us restless and discontent, and usually unable to enjoy life.

There is a rule that can be of great help here, not only when we are working, but in each moment: "Do and say nothing unnecessary." Asking yourself what is truly necessary can make a huge difference in your life. Ask it in all kinds of circumstances – when you are tempted to criticize or gossip, but also when a friend is silently crying out for something and you have not noticed because your attention is on yourself.

I like to keep in mind the four laws of ecology that Barry Commoner shared with the world in *The Closing Circle*:

1. Everything is connected to everything else.
2. Everything must go somewhere.
3. Nature knows best.
4. There's no such thing as a free lunch.

One of the remarkable aspects of these four principles is that they are all saying the same thing. Everything that you think, do, or say is connected not only to everything else in your life, but to everything in everybody else's life too. You cannot do anything in isolation. Also, each person is connected to everyone else. Looked at this way, it becomes evident that there is only one of us. Buddhists explain this by saying that each of us is like a different part of the same body, and yet we strut about, believing that we have a life of our own, unaware that we may be just a cell or an eyelash.

The last word on practicing sustainability goes to May Sarton, who, in *Plant Dreaming Deep*, brings home the way to make it truly personal: "Experience is the fuel; I would live my life burning it up as I go along, so that at the end nothing is left unused, so that every piece of it has been consumed in the work."

**

Toinette Lippe *worked in book publishing for 50 years and now devotes herself to East Asian brush painting. She lives in New York City.*

Acknowledgment This essay is recycled from material in *Nothing Left Over: A Plain and Simple Life*. Copyright © 2002 by Toinette Lippe.

Chapter 33

Teaching Sustainability in the Anthropocene Era

Kai Lee and Richard Howarth

In our roles as teachers in environmental studies programs at liberal arts colleges, we have mentored many young people whose abiding concern for the environment and for social justice drives a deep desire to live in a manner that lives up to the ideals of the newfound "sustainability" movement. Indeed, we share these values ourselves, doing our best to do our parts as good citizens who think globally and act locally. Still, we know all too well that one individual's decision to drive a hybrid, eat low on the food chain, or recycle a beverage container is like one drop of water in the sea of forces that are pushing our world in the direction of a socially unjust, environmentally unsustainable future.

We of course seek to nurture our students' sense of commitment and their attempt to reconcile their lifestyle decisions with their personal values. At the same time, we worry that the choice to become a "green" consumer is no substitute for the deliberative, transformative engagement that our students will need to achieve to realize their shared vision. Sustainability problems, we teach to our students, result from political failures and require political solutions. In a society so strongly dominated by its emphasis on individual identity, consumerism, and the magic of the marketplace, the idea that personal actions to achieve sustainability are not the alpha and the omega can seem alien to many "sustainability" advocates, who hope that casting dollar votes in the

> We know all too well that one individual's decision to drive a hybrid, eat low on the food chain, or recycle a beverage container is like one drop of water in the sea of forces that are pushing our world in the direction of a socially unjust, environmentally unsustainable future.

marketplace will make up for the perceived ineffectiveness of politicians in Washington or Brussels.

We live in a time that some are calling the Anthropocene Era – an age of human domination in which our species is propelling change on a scale comparable to the forces of nature. The sweeping land-use alterations of agriculture and urbanization – coupled with the introduction and extirpation of species – are driving the loss of biodiversity and altering the ecosystems of rivers and coastal regions. Climate change, caused by the burning of fossil fuels, is already transforming the polar region and its peoples and is shifting growing seasons, the frequency of droughts, floods, severe storms, and the chemistry of the oceans, in addition to raising temperatures. The expanding human footprint of the past three centuries marks a path that has become plainly unsustainable. This is an inconvenient truth, but one that can no longer be ignored in many places and in increasingly many industries.

Moving toward a sustainable relationship between humans and natural systems will require people to act differently, which will in turn require, over time, major alterations of the institutions that shape human behavior. Understanding human activities and institutions – the objective of the social sciences – is accordingly essential. Yet attention to the human dimensions of environmental change remains inconstant and superficial, and efforts to improve the relationship of people to nature are entangled in ideological conflict.

If a landowner preserves a forest, he benefits those downhill and down the watershed from his property, who face lowered risk of floods and have more clean water. But the upstream landowner can usually make a lot from cutting down the trees and using the land to grow crops or to build cities. Usually, too, the forest owner cannot collect any revenues from the downstream beneficiaries of the forest. So the trees often get cut, inflicting harm to land, waters, and communities and increasing the burden of greenhouse gases in the atmosphere. This is a story of systemic dysfunction: No one is *intending* to wreak damage on the natural world, but that is the result.

All nations are deeply committed to a global economic system that yields unsustainable outcomes. We are in the position of the forest landowner, whose expectations are tied to economic growth and high revenues even when irreversible changes to nature result. Thus, the practical goal of sustainable development is to move *toward* greater sustainability, rather than to achieve an ideal end state. Indeed, it is far from clear that a human population approaching 10 billion can live sustainably with only the technological capacities currently known and the institutional arrangements now in place.

Three global-scale elements of the Anthropocene are accelerating – climate change, erosion of biodiversity, and urbanization. Each of these

> Sustainability problems result from political failures and require political solutions.

poses a grand challenge: Science can measure the alterations of social, biological, and physical processes under way, but ways of instituting an effective management of those changes are not known.

There are hopeful developments in many areas nonetheless. The efficiency of energy use is rising, so that more light, transportation, and the other desirable outcomes of energy use are being produced for each unit of energy consumed. In the U.S., sales of gasoline may have already passed an all-time high – a trend that will increase further with rising oil prices. Energy use per person in the state of California has been roughly constant for a generation, even as its economy has grown enormously.

Over the past 30 years, increasing areas of land and sea have been classified as protected areas, in an attempt to maintain valuable, biologically diverse ecosystems. Building construction in industrialized economies has been moving toward high-efficiency "green" designs; innovations in mass transit are making their mark in developing countries such as Mexico. Consumers and retailers are building demand for products that are produced more sustainably, such as organic vegetables and sustainably harvested seafood. None of these innovations, by itself, has reversed the unsustainable trends within its economic sector, but there are active ferment and a search for solutions that can scale up to have a significant impact.

The central puzzle is the governance of commons: resources that are shared by many stakeholders, none of whom has both the incentive and the capacity to manage them well. Fish in the ocean become valuable, economically, when they are caught, but often no one can take responsibility to see that fish populations are maintained at levels where the catch benefits humans in a durable fashion. The key verb in that sentence is "can": The problem with commons is not an individual's failure to exercise responsibility, but rather a collective problem rooted in inadequate institutions.

One answer is managing fish harvests through a system called "catch-shares." Groups of fishermen are given the legal right to catch a specific percentage of the total harvest of a species. The total harvest is set using scientific estimates, using data from the fishers themselves as they sell their catch. The owners of the catch-shares can buy and sell the right to gather a portion of the harvest. Over time, an efficient fishing fleet emerges, as the costs of boats, fuel, nets, and other gear sort out, and rewards the efficient harvesters. Catch-shares are used in New Zealand and increasingly many other places in the world. Such a reorganization of human activities moves an industry toward a more sustainable profile.

The World Wildlife Fund is now working with major corporations to improve the security of those firms' long-term supplies of resources. Coca-Cola needs clean water for its bottling plants; Home Depot needs lumber to sell from responsibly managed forests. These firms' motives are not altruistic, yet they lead major economic actors toward more sustainable outcomes. If these activities can become part of the normal course of business – the way that buying insurance against risks has become routine – it is possible to see how significant improvements in sustainability can come to a whole economy.

The importance of social process and institutional design is gaining traction. The recognition of Elinor Ostrom's work on common-pool resources with the 2009 Nobel Prize in Economics may be the most visible academic statement.

The environmentalist and poet Wendell Berry wrote, "Our understandable wish to preserve the planet must somehow be reduced to the scale of our competence." But the grand challenges of sustainable development point in a direction opposite to this small-is-beautiful thinking. Rather, the task is to enlarge the scale of our competence, so that the world-changing forces of the Anthropocene can be domesticated.

Make no mistake: This will be a long haul, riddled with errors and hard-earned lessons. Consider environmental regulation, which has improved public health and environmental quality substantially. Yet those benefits have been won by means that are bureaucratic, sometimes inefficient, clouded by scientific uncertainty, and often reviled by the industries bearing the burdens of regulation. They are like the forest owner: paying the costs of controlling pollution but unable to capture the benefits. The result has been a generation-long struggle to devise regulatory methods that can harness the self-interest and agility of innovative businesses to achieve environmental goals.

> Moving toward a sustainable relationship between humans and natural systems will require people to act differently, which will in turn require, over time, major alterations of the institutions that shape human behavior. Understanding human activities and institutions – the objective of the social sciences – is accordingly essential.

The cap-and-trade mechanism used in the U.S. to regulate the emission of sulfur dioxide is one successful experiment. As with catch-shares, the right to emit sulfur dioxide has been parceled out to individual firms; the total quantity allowed in a given year is set by the Environmental Protection Agency. Firms all over the country can sell their rights, if they run a low-polluting business, while others buy equipment to reduce their emissions – or buy additional rights so they can emit more. These are business decisions, akin to buying crop insurance or advertising, and businesspeople are used to making such choices in ways that allow them to make money. In the process, emissions nationwide are lowered over time by the government, at a far lower cost than if each firm had to conform to the edict of agencies that cannot see the trade-offs facing the individual companies.

Environmentalism as a social movement has been enormously influential. Most nations now argue over environmental protection as significant elements of their domestic policy. Most major corporations and international institutions take environmental variables into account in their decision making. Environmentalists have demonstrated their scale of competence includes the conservation of precious local places and species such as the bald eagle. Environmental advocates are significant critics, and sometimes partners, of businesses and governments as they wrestle with environmental problems. These changes in the way decisions are made are movements toward greater sustainability, even if the choices that result do not always go that way.

Over the past half-century, individuals and organizations have moved toward sustainability via politics, markets, shifts in cultural values, and through actions guided by the conscience of billions. The grand challenges of climate, biodiversity, urbanization, and sustainable development have yet to be fully met. While technological innovation is obviously important, so too is the harnessing of a growing understanding of human behavior and institutional design. The creation of coupled human and natural systems that are durable and vital in their dynamism has begun, but there is much more to do. In an important sense, humans are the most important untamed species on the planet, but the process of domestication is under way.

Sustainability is not a concrete destination, nor is it a property of goods and services that individuals can effectively promote via the marketplace by adopting a "sustainable" lifestyle or consumer identity. Rather, it is more like freedom or justice, a direction in which we strive, along which we search for a life good enough to warrant the comforts we enjoy. Freedom and justice are often taken for granted, even though many have suffered and died in their pursuit and defense. A materialist condition like sustainability is harder to imbue with romance. But the enormous changes of the Anthropocene leave us a difficult choice: to accept our humanity in the company of the whole human race and the natural world we jointly share, or to concede that being human is too difficult for the richest, most advanced beings in history.

***

Kai Lee *is program officer for science at the David & Lucile Packard Foundation and Rosenburg Professor of Environmental Studies, emeritus, Williams College.*

Richard Howarth *is the Pat and John Rosenwald Professor of ecological economics at Dartmouth College, and editor-in-chief of the journal* Ecological Economics. *Lee and Howarth are co-authors, with the late William Freudenburg, of* Humans in the Landscape.

Chapter 34
Don't Sustain; Advance

Kevin Finneran

Webster's Dictionary defines boring as "an essay that begins with a definition from *Webster's Dictionary*." Not really, but it is a common way to begin an essay on sustainability because writers on the topic are like Humpty Dumpty in *Through the Looking Glass*, who tells Alice, "When I use a word, it means just what I choose it to mean – neither more nor less." If we do turn to the dictionary, we find that the word "sustain" can mean to provide with relief, to nourish, to prolong, to support, to buoy up, to bear up, to confirm. All of these definitions imply a goal of preventing matters from getting worse. What is curiously missing is any hint of progress, innovation, or the creation of something better – and therein lies the problem with sustainability.

In practice, "sustainability" and its cousin "sustainable development" can be used to mean everything from returning to a past Eden that exists in the amber of someone's imagination to preserving as much as possible of the current state of the natural world and its resources. The difficulty arises when this sense of preserving is extended to the social world where the unintended outcome of the call for sustainability is that it can lead to the rich preserving what they value in the natural world and the poor enduring a life with too little access to the comforts provided by exploiting the planet's resources.

> Sustaining implies preserving and maintaining, whereas development demands change. Therefore, the concept of sustainable development has an inherent tension that's difficult to reconcile.

What is curious is that although the flacks for sustainability harbor such vastly different notions of what the term means, they want us to believe that they are all somehow on the same page. What a vast, wrinkled, blotted page it must be. Or perhaps they have smoothed it all into an elegant M. C. Escher drawing in which water flows up and down at the same time. In any case, it incorporates some form of optical or intellectual illusion because it certainly can never satisfy the desires of all who march behind its banner.

The inherent contradiction is that it wants to preserve a vision of the natural world in which humans are not driving the bus, to protect specific parts of the natural world that the wealthy enjoy seeing and visiting, and to achieve this while also providing a dramatically enhanced standard of living for the planet's billions of impoverished people. It wants to stop the momentum of the technological and industrial beast that has provided them with the luxury to think about sustainability. They fear that the billions of people in the developing world will do as we do, not as we say, and follow in our footsteps to prosperity by tapping the Earth's resources even it means altering the state of the land, the water, the air, the flora, and the fauna.

The challenge for the advocates of sustainability is to convince the industries that have made us rich and comfortable and the people who have benefitted from resource-fed productivity that they must behave differently while reassuring the legions of the poor that there is a better, though untested, path to prosperity. While coal powers our air conditioners and oil propels our cars, we lecture them about the virtues of wind generators and photovoltaic cells, overlooking the harsh reality that power from these sources is much more expensive. Yes, we've claimed the most fertile lands and supercharged their productivity with fertilizers, pesticides, and irrigation, but proponents of sustainable agriculture urge the developing countries to use their creativity to find ways to boost productivity without these aids – and no cheating with those genetic engineering shortcuts. Better to starve than to risk offending Mother Nature.

To be fair, not all proponents of sustainability use the term to represent a rigid set of rules and principles. The most sensible and flexible definition of sustainability is the one offered in *Our Common Future*: "In meeting our needs today we should avoid doing anything that would limit the ability of future generation to meet their needs." This includes no specific commandments such as "Thou shalt not use pesticides" or "Thou shalt immediately replace

> A desire to "have it all" will prevent us from facing decisions about unavoidable trade-offs. I find it impossible from my perch of privilege made possible by development and industrialization to preach to desperate people about the necessity for them to do things differently.

fossil fuels with renewable energy sources." However, it assumes that we can know what tools and skills will be available in the future. Although it might seem that using oil now will make it unavailable in the future, we do not know how much oil is available or how much will be needed in the future. We can find new ways of extracting oil, technologies that use fossil energy more efficiently, and alternative energy sources that make oil less necessary. Extracting water from aquifers at a rate above replenishment doesn't sound like a good idea, but we cannot predict what low-cost desalinization and extensive water distribution systems could mean for the world's water supply.

Indeed, sustainability is a backward-gazing stance that worries that if we keep doing what we've been doing with the technology and resources that we have, there will be trouble. But if we extrapolate from almost any point in history, it is clear that continuing to do what we were doing with what we knew would have led to disaster in a few hundred years. But throughout history we have learned to use resources more productively and to find new resources. Science and technology have made it possible to support unsustainable population after unsustainable population.

The dilemma facing believers in sustainability is that there are billions of impoverished people on the planet who are understandably not content with sustaining anything. They want more of what the rich have, and they see no way of achieving that goal other than doing what the rich did to acquire what they have. Of course, the faithful recognize this and argue that to achieve their dream, the rich will have to be satisfied with less so that the poor can have more. There's the rub. They have no power over their rich neighbors, and the rich are not volunteering in large numbers to become less rich. What they want to sustain is their quality of life and their place in the world. But the idealists of the wealthy nations do have power over the poor, which they can exercise through national development assistance programs, international organizations such as the World Bank, and a proliferating number of nongovernmental organizations. They can lean on them to sustain forests, eschew industrial agriculture, employ renewable energy sources, and preserve traditional practices.

Preventing matters from getting worse is not enough when the lives of billions are miserable in ways that are inconceivable to inhabitants of the industrialized world. These billions find it hard to imagine a world that is worse. They want a world that is dramatically richer and better. They want to squeeze more from the Earth's resources before they meet their untimely deaths from hunger and disease. If the rich want to preserve the rainforests, they can buy them; if there are oil and gas to be recovered, get them; if biotechnology can customize crops to enable to thrive with limited rainfall and poor soils, what are we waiting for?

In a dream world, miraculous innovations (but not industrial agriculture, genetic engineering, nuclear power, tar sands, fracking, improved internal combustion engines, or a host of other technologies that actually work) will generate more from less and provide the only long-term path out of poverty. It's a beautiful dream. But when that dream fails, the rich can continue to benefit from the wealth they have extracted, and the poor will be worse off than ever. The vision of sustainability is an indulgence that only the rich can afford. Until the advocates of sustainability transform the industrialized world, they should refrain from lecturing those who desperately want the reality of industry and technology rather than the promise of sustainability.

> We should move away from the theme of preserving and protecting with its hint of nostalgia for a simpler world and embrace the idea that change and innovation, albeit with a consideration of limited natural resources, is the only path that can improve the lot of the impoverished billions.

Envisioning a world that uses resources more sparingly and produces goods with minimal environmental impact is a noble quest, but in the meantime we need to use the technology we have and the resources we can find to improve the lot of the billions who need more than an idealistic vision. Our focus should not be on preserving, but on improving, expanding, and innovating for the benefit of those who have little of value to preserve.

***

Kevin Finneran *is editor-in-chief of* Issues in Science and Technology. *He is also director of the Committee on Science, Engineering, and Public Policy, a joint unit of the National Academy of Sciences, National Academy of Engineering, and the Institute of Medicine in Washington, DC. Earlier he was Washington editor of* High Technology *magazine, a correspondent for the* Financial Times *energy newsletters, and a consultant on science and technology policy. He is a fellow of the American Association for the Advancement of Science.*

Chapter 35

Changemakers for Sustainability

Karabi Acharya

I am an urban creature. Never lived on a farm. Never really learned the difference between a cow and a steer until I was 25. I always feel nervous when out of cell phone range. But I am an "engaged" urban creature. I've read Michael Pollan, watched *Food, Inc.*, and diligently purchased CSA shares. Of course, I have a small herb garden and an occasional tomato plant. I understand the consequences of mono-culture farming and the damage caused by pesticides, the tragedy of deforestation across the globe. My children only drink organic milk and themselves prefer Starbucks to McDonald's. And now I know that I was just fitting into some preconceived notion of what I need to know and do about the sustainability of our food system. I was the Pottery Barn version of "concerned urban creature!"

And one day a couple years back, I went to a real farm in Vermont, an organic farm and dairy. I had gone for workshops on sustainability. We had rich discussions and thought-provoking experiential exercises. Yet none of this came close to what I learned from the food I ate during those workshops. Each day, we savored homemade, breads, cheese, tomatoes, and beets picked that morning. We ate food of every cuisine, from Thai and Indian curries to squash pies and pasta. Members of the community lovingly prepared each item (many of them very time-consuming) down to the cookies and cereal bars offered as snacks. I knew the cow who gave us

> Sustainability is about this feeling in your gut, your heart, your soul, that the earth, the soil, and every human being are deeply connected to one another; that we are part of some super-organism; and that creates a reinforcing virtuous cycle.

milk by name. I knew which row the zucchini came from. And in slow motion, as if for the first time, I *felt* a connection to the soil, to the plants, to the cow, to the milk, and to the people who prepared it all. I felt it in my soul that the life of all these beings and the soil upon which we all grow are deeply connected. I found myself becoming indignant when the smallest morsel of food was wasted – a strange feeling from someone who grew up "feeling bad" about wasting food. This was different – I mean the wasted energy that went into cooking and growing it all; the waste and disrespect for the people who poured attention, and even love into this food. I got it. I moved from knowing about food systems to feeling food systems.

Sustainability – of food systems or any other type of system – is not some abstract concept. It's not a new role we take on in our lives (like the latest trends) – installing CFLs and recycling. It is certainly not about buying different stuff – a Prius, organic food. It is so easy to keep sustainability in your head, to intellectualize it. We can draw complex systems diagrams, do economic analyses, and make projections. But I understand now that sustainability is about this feeling in your gut, your heart, your soul, that the earth, the soil, and every human being is deeply connected to one another. I have found that the way to understand sustainability is through the heart, not the head. And I ended up learning one of the most essential lessons from the founder of this farm, Donnella Meadows:

> It is not easy to practice love, friendship, generosity, understanding, or solidarity within a system whose rules, goals, and information streams are geared for lesser human qualities. But we try, and we urge you to try. Be patient with yourself and others as you and they confront the difficulty of a changing world. (From *Limits to Growth*)

So what is my vision of a world that practices "love, friendship, generosity, understanding, or solidarity" and I would add empathy and change-making to that list?

What if we took this seriously? What if schools were required to teach and facilitate these practices and of course measure progress on them just like reading and math? (Don't let anyone tell you there is no way to find proxy measures for all these concepts.) What if all workplaces fostered an environment that promoted these practices? What if there were a "currency" for empathy and social capital? What if there was a way for people to experience being blind and learn about the "blindness" in all of us? These are not simply pipe dreams; Ashoka Fellows are creating this world. Eric Dawson with PeaceFirst not

> Sustainability as a concept is compelling in that it is timeless. It is about creating virtuous and renewable cycles.

only teaches children to practice friendship, empathy, and changemaking but supports changes to the school environment for these practices. Caroline Casey of Kanchi is changing the culture of workplaces to celebrate people's unique abilities rather than the "disabilities" focused on by the larger society. TimeBanks is a way for community members to exchange their time and skills, the "currency of empathy" as its founder Edgar Cahn calls it. And Andres Heineke created Dialogue in the Dark as a way for all of us to empathize with the blindness in all of us.

As Bill Drayton, founder of Ashoka, describes, in the past people could turn to rules and hierarchies for guidance, but the world is changing too fast now. Hierarchies are flattening, rules are constantly in flux and often in conflict, and information flow is rapid and ambiguous. As the rate of change increases, every person needs an ever-higher level of empathetic skill in order to thrive. Where rules are no longer clear, without the skill of empathy, we will hurt people and disrupt institutions. Conscious empathy at a high level has become essential to being able to participate fully in a changing society and to act and contribute positively in the world. Ashoka envisions an "everyone a changemaker" world, where every member of society has the freedom, confidence, and societal support to address any social need. Such a global society will foster innovation and the desire for change, so that individuals will find within themselves the potential to make change.

> Too often, I feel sustainability is "owed" by "environmentalists," which can feel exclusive. We talk so much about the ecosystems involved. But at its heart, sustainability is about people; people caring enough to act. We need to talk more about the people skills needed.

For 30 years, Ashoka has been building a network of more than 3,000 leading social entrepreneurs around the globe bringing systemic change for the good of all in every area of need. Being at the center of this network provides us a deep understanding of the key levers for bringing about structural social change in society, across industries and sectors. We see patterns that show where interventions are most needed in society and where fields are ripe for change. We then align the key players in collaborative entrepreneurship to accelerate that change. Ashoka's influence in my own search for understanding and for solutions has been profound.

We each could decide to take very seriously the cultivation of love, friendship, generosity, understanding, solidarity, empathy, and changemaking in ourselves, our children, our colleagues, our friends. Go ahead and change your lightbulbs, buy better (and less) stuff, but the strongest lever for change that we have for sustainability is people; people's deep connections to other people, to the planet and the recognition that they are part of the system and thus have the power to change it. As for me, I've expanded my garden beyond herbs, I still buy organic milk, but I am now completely focused on how to make two young girls into the best-loving changemakers they can be.

***

Karabi Acharya *is the global director of Ashoka Impact. She leads efforts to understand how Ashoka Fellows change systems. For over 15 years, she has worked in global health and development in over 12 countries. She has a degree in anthropology and a Doctor of Science from Johns Hopkins School of Hygiene and Public Health and is a Donella Meadows Leadership Fellow. Acharya welcomes crossing boundaries and steadfastly refuses to choose one professional label or work in one sector and is currently focused on cultivating more changemakers in the world.*

Chapter 36
What Social Entrepreneurs Taught Me About Sustainability

Mirjam Schöning

When the Schwab Foundation for Social Entrepreneurship set out to search for leading social entrepreneurs in the year 2000, our first task was to define what we were looking for and set up clear criteria. While the phenomenon was surely not new, the term "social entrepreneur" only began to catch on at the beginning of the new millennium.

At first, social entrepreneurship was in the shadows of the "dot-com" boom. I moved to Boston in 1998 for an MPA program at the Harvard Kennedy School. A friend of mine invited me to her housewarming party. She had taken over the room from a guy who had started an Internet company in his first year as a Harvard MBA student and sold it for several million upon graduation. Entrepreneurship was clearly *en vogue* and most courses at the business school had "entrepreneurial" in their title: "Entrepreneurial Finance," "Entrepreneurial Marketing," and so on, luring many who wanted to learn about making a quick buck.

> Sustainability means ensuring long-term survival by taking social, environmental, financial, economic, and political factors into account.

In this new gold rush, I was particularly drawn to a different entrepreneurship course called "Entrepreneurship in the Social Sector," one of the first courses focusing on social entrepreneurship. A year later, as the dot-com boom crashed in 2001, social entrepreneurship came out of the shadows and has seen an exponential rise in student interest, not just in Harvard, but at most major business schools around the world.

Sustainability might well be an important underlying factor for why social entrepreneurship has been so appealing. Not only did the crises in

2001 and 2008 show that purely profit-driven ventures can be ephemeral. In addition, there is an increased desire to find long-term solutions to the world's social and ecological problems. The attention has shifted to highlight and encourage entrepreneurs who implement innovative methods to tackle poverty, health, education, and/or environmental challenges through market-based solutions.

Muhamad Yunus, the founder of the Grameen Bank and Nobel Peace Prize recipient in 2006, became the most well-known example of a social entrepreneur. He showed that you can lift people out of poverty by providing microloans and savings instruments that paid for themselves. It was these types of examples we were looking for at the outset of the Schwab Foundation. The idea of the founders, Klaus and Hilde Schwab, was to ensure a much larger visibility for the innovative methods of social entrepreneurs and help disseminate them through the platforms at the regional, global, and industry levels of the Foundation's sister organization, the World Economic Forum, which was also founded by Professor Schwab in 1970 under the notion of "entrepreneurship in the public interest."

> Isn't the concept of sustainability engrained in human nature?

While we knew what a social entrepreneur was when we saw one, no clear definition existed at the time. The first academic conference on social entrepreneurship made this painfully visible when academics fought for two days and the only common denominator at the end of the debate was that social entrepreneurs bring about "social change."

When we defined the criteria for the leading social entrepreneurs we were looking for, there was no doubt in our minds that "sustainability" is one of two key criteria, next to "innovation," or the fact that a social entrepreneur brings about social change by transforming traditional practice through a new service, product, or approach.

For the selection of leading social entrepreneurs, it was important to assess that both the impact of the social enterprise is sustainable as well as the organization itself. Many social entrepreneurs directly pioneer solutions that offer a more sustainable future.

Reed Paget is the founder of Belu Water, a bottled drinking water company in the UK. It was the first to use compostable bottles made from corn instead of PET bottles, triggering some of the industry's giants like Coca-Cola to follow suit with its plant bottle. It also was the first bottled-water company to become carbon-neutral and donates its profits to clean water projects in Africa and South Asia.

Belu's impact clearly is a role model for environmental sustainability. In the earlier years, it was less obvious if the organization itself would be sustainable. As an indicator to assess if a model can be financially or

> I would embed sustainability in school curricula to teach, from first grade on, that financial, human, and environmental resources are limited.

economically sustainable, we assess the revenue stream of an organization. A social enterprise should be able to recover a fee for its products or service, at least partially. This may come from the public sector, particularly if it is for a public service such as primary education or basic healthcare.

Lately, there is a particular excitement around those social businesses that are able to break even or even run profitably while providing a significant social or environmental solution. After many initial struggles with investors, Belu Water has reached that level. These social businesses have the advantage that they do not need to rely on fundraising and donations, which typically frees up between 30 and 50% of the founder's time. It is therefore not surprising that we have observed much faster growth rates among the financially self-sustainable social businesses compared with the more nonprofit-oriented ones.

However, it would be an illusion to only consider profitable social businesses as sustainable in organizational terms. Newspapers are full of previously highly profitable companies

What to do:
- Follow your passion – it's the most important ingredient and it's the one thing that will keep you going when the going gets tough.
- Balance passion with rationale – Are you truly addressing a real need? Can this be backed with facts and figures?
- Carefully choose your business model – try to achieve revenues from day 1.
- Build evaluation and impact measurement into your processes from the beginning.
- Consider a **"social franchise"** – what we arguably need most today are entrepreneurial people that take up a brilliant model from one part of the world and implement and adapt it in another.

What not to do:
- Don't be afraid of generating 1,000 ideas – and dismissing 999 of them, and to let new ones emerge.
- Don't give up before 36 months. Invariably, this seems to be the time it takes to get a social enterprise off the ground.
- Don't just assume your idea is unique. Study approaches that lead to the same impact you are trying to achieve. Is your approach really as unique as you think it is? Are there more proven methods to achieve the same outcome from which you can learn?

> **BioRegional: An Example of a Sustainable Social Enterprise**
>
> Under a nonprofit umbrella, BioRegional develops business solutions following the 10 **"One Planet Living"** principles:
>
> 1. Zero emmissions
> 2. Zero waste
> 3. Sustainable transport
> 4. Sustainable materials
> 5. Local and sustainable food
> 6. Sustainable water
> 7. Land use and wildlife
> 8. Culture and heritage
> 9. Equity and local economy
> 10. Health and happiness
>
> Examples for BioRegional projects include several housing and office units across the world, a barbecue charcoal production company in the UK using sustainably managed wood from southeast England to replace the 98% of charcoal imported from often unsustainably managed sources, and a a recycling company to meet the UK's needs for paper.

going into bankruptcy. Social enterprises can also deliver a sustainable or lasting impact if the organization relies on grants. Consider KickStart, which has developed pumps and other technologies for poor farmers in Kenya and other African countries, enabling them to establish small, profitable enterprises. While the R&D and initial marketing costs have to be written off, the farmers are able to pay a price that covers the manufacturing, sales through local merchants, and after-sales costs of the pumps. The micro-irrigation pumps typically enable a farmer to grow fruits and vegetables throughout the year and sell them in the market. They recover the investment in 3 months and start making an average profit of US$1,100 per year. Thus, a self-sustainable economic cycle can be "kickstarted" with smart, time-bound subsidies to develop and market a product for the poor and create a sustainable social impact.

Finally, we have seen the influence of the political environment in which a social entrepreneur operates on the long-term viability of the organization and its impact. The microfinance sector in Andhra Pradesh, India, has severely suffered largely due to the state's decision to introduce adverse regulation, such as caps on interest rates and monthly repayment schedules. Meanwhile, the Bangladeshi government has forced Yunus to step down as managing director of the Grameen Bank. A change in leadership might not threaten the sustainability of the bank and its impact, but there is a danger that the government could use the bank as an instrument in elections by providing loan forgiveness or handouts, which would essentially bring the repayment morale to an end.

Social entrepreneurs address many dimensions of sustainability. While I previously associated the term primarily with ecological sustainability, the social enterprises in the Schwab Foundation network taught me the importance of considering other aspects of sustainability: the social, financial, and political dimensions.

***

Mirjam Schöning *is head and senior director of the Schwab Foundation for Social Entrepreneurship in Geneva, Switzerland. Schöning holds an MBA from the University of St. Gallen, Switzerland, and studied international business at ESADE, Spain, and the Stockholm School of Economics, Sweden. She graduated with a Master in Public Administration from the Harvard Kennedy School of Government. Previously, she was a consultant at Bain & Company. She analyzed the social sector lending strategies in Latin America for the World Bank, and worked on a regional centralization strategy for Shell in Scandinavia.*

Chapter 37
An Emotional Connection with Sustainability Through Documentary Films

Heather MacAndrew and David Springbett

For over 35 years, we have been independent documentary filmmakers. By our last count, we've filmed in 32 countries around the world – including our own, Canada – that were grappling with issues of what we could now call "sustainability." What we have seen has taught us a lot and, inevitably, raised more questions. No one size fits all; people's lives are complicated, no matter where you are. The "simple life" doesn't exist.

Because making a documentary film usually involves talking with a whole range of people – from theorists to practitioners, from slum-dwellers to presidents – we tend to get a wide range of perspectives. We believe you have to ask tough and often uncomfortable questions to try and understand why a given situation is the way it is and how it can be made better. This hopefully leads us and our audiences to think about problems in different ways.

In our early days, we were interested in calling attention to voices, from what was then called the "Third World," not often heard in mainstream media. If, as we believed, "bottom-up" development was important, then we wanted to find out what "the bottom" was thinking about. At that time, the term "sustainability" did not exist in the mainstream vocabulary. So before the word "sustainability" was even coined, how did we identify the concept through our film projects? What did we see?

> To us, sustainability has to do with old teachings: Do no harm; treat the planet as you would want to be treated; do not take more from the earth than your share. But if you're poor and desperate, survival nearly always trumps sustainability. For the privileged, like us, it means changing profligate habits and using just what we need – not what we want.

Looking back, for us "sustainability" might have been homes being rebuilt after an earthquake, with bricks made from locally-available mud and straw. Terraced fields, whose crops fed the animals, which in turn fertilized those fields. Tin cans, tires, scrapped car parts, being repurposed in a thousand different ways. Girls and women becoming literate, learning about disease prevention and how to avoid unwanted pregnancies. Tying it all together, we've seen people with a sense of community, and ideas of what *they* perceived they needed to make their lives better.

We've also seen the "unsustainable" in the form of imported farm machinery rusting in fields; a makeshift warehouse full of donated cans of spinach sitting untouched a year after an earthquake; factories idle and collapsing; health teams unable to travel to immunization clinics for want of fuel for their vehicles.

We never set out to make films about sustainability, but when we look at our work through a sustainability lens, we see that the real lesson we've learned is that everything – the social, environmental, and political – is connected. You can't change one thing without affecting everything else.

Here are a few examples of what some of our films over the years have shown us:

"Guatemala: Campo Vivo" was a half-hour documentary we produced with the Canadian Broadcasting Corporation in the aftermath of the devastating Guatemalan earthquake of February 7, 1976. The film explored different impacts of short-term aid and long-term development; community leaders in one Mayan village understood and expressed the difference, in their own words. It was our first and in some ways most important lesson in sustainability. And it led us to think about a difficult question: What is the connection between inequality and sustainability? What is the relationship between political repression and inequality?

Research took us to a highland Mayan village that had been leveled by the earthquake. For over 10 years prior to the quake, the village had participated in an agricultural development project supported by an Oklahoma-based NGO, World Neighbors. The project had worked according to some basic principles: bottom-up, not top-down; each one, teach one; a field becomes a classroom and the village a school; every farmer has an "experimental farm" – sometimes only a meter square, but nevertheless some place to learn better ways to grow crops on marginal land. The results were impressive. For the first time the community was virtually self-sufficient in food. For them, the earthquake was just one more impediment facing people who were used to working together to overcome obstacles.

What did we discover? That learning how to grow more food for your family is a basis of rural sustainability; receiving cans of spinach in food aid is not. We learned what made the World Neighbors project work well, but we also learned about larger social forces that could attempt to destroy not just one village but a whole people.

Mayan farmers, having only small parcels of land, often worked as transient laborers on the coffee, sugar, and fruit plantations. In the village we filmed, as people began to grow enough food for their families, fewer farmers needed to go away for months at a time to work on plantations. If villages became "sustainable," was there a threat that cheap labor for the plantations might disappear? Was "sustainability" therefore a threat to those with power? Difficult questions.

> While the concept of sustainability was orginally appealing – who could argue with such a motherhood concept? – we now find we're conflicted. "Sustainability" has come to mean almost anything anyone – from ad executives to government spin doctors – wants it to mean, to serve their own purposes.

How do we become the people we are? How does *where* you grow up influence *how* you grow up? That was the central question of our 1986 film, *Growing Up in the World Next Door*. The film profiled three young adults in three developing countries: Nepal, St. Vincent, and Kenya. We had first filmed these three individuals in 1979–1980 when they were all about 12 years old; now we set out to record the changes 6 years had made in their lives. All three of our subjects had interesting stories, and their lives were in large part defined by the social and economic structures of their societies; but the one whose story is perhaps most relevant to the question of sustainability is Bikas, who lived in Nepal. His story turned out to be a surprise. He and his family lived in the village of Tupche on the Trisuli River, a slow 6-h drive from Kathmandu. In the late 1970s, the village had participated in an integrated rural development project funded by the World Bank. Bikas's family, along with others, had benefited. Their crop yields improved, they were able to buy more land, and they expanded their house, turning it into the village hotel.

In 1985, when we came back to make the follow-up film, Bikas, the eldest son, was going to high school in Kathmandu. The family's plan was for Bikas to go on to college, get a job, and then bring the whole family into Kathmandu to live. So the irony was that a "model" development project designed to make rural life better turned out – for at least one family – to be a ticket to the city and the enticing possibility of a middle-class life. How do you keep 'em down on the (sustainable) farm once they've seen Kathmandu? If improved rural life didn't keep them on the farm, was it truly sustainable? If younger generations continue to leave their rural homes for cities, does rural life have any hope of being sustainable?

In Kenya, Michael, 1 of 16 children (two mothers), lived in a drought-prone area near the Somali border. In 1986 he was making his way through high school, as funds for tuition allowed and when drought didn't decimate the family's crops. There were still lots of "ifs" about his future.

Patsy lived in a tiny village on the leeward side of St. Vincent in the southeastern Caribbean. She had finished high school and was looking for work, living at home, and packing bananas as a casual laborer. But shortly after the film shoot, she became pregnant and began life as a single mother, eventually leaving St. Vincent to find work on another island.

At the end of *Growing Up in the World Next Door*, the narrator muses about the three young people whose lives we have just seen: "*These three want what everybody else wants: a job, a place to live, families of their own. What will help them reach their heart's desires?*" What, indeed, as long as the world economic order continues to favor us in the North, with our mostly unsustainable lives? That prompts another question: Is an economic system that uses "growth" as its measure of success itself sustainable?

In a world where wood and wood products are used virtually everywhere, and industrial forestry is the norm, is it possible to have jobs *and* trees? What would it take to have a sustainable forestry system *and* create steady jobs for people? These were the basic questions we wanted to explore in our film *GoodWood*, made in 1998, for the long-running CBC series *The Nature of Things, with David Suzuki*.

> We would encourage people to ask basic and essential questions: What do I really mean by sustainability? Is my action or practice contributing (or not contributing) to sustainable living? We'd ask people to challenge assumptions: If sustainability is such a great idea, why is it so hard to put into practice?

In a tiny indigenous Peche community in Honduras, a couple of American woodworkers wondered what they might do to encourage the use of other, not endangered local hardwoods and at the same time help create a community business in an impoverished place with few options. Working through an organization, GreenWood, whose initial goal was convincing woodworkers in the U.S. and Canada to also use LDS (less developed species) they worked with the Peche community to develop a cottage industry: bentwood chair-making. The chairs were made with simple tools and used what had previously been considered "garbage wood" rather than purpose-cut endangered species and were sold in nearby towns.

The film went beyond that small village micro-enterprise by looking at a community-owned mill in a Zapotec community in Oxaca, Mexico; and on a much larger scale, at a plywood mill in Oregon that used only wood from its own sustainably harvested forests.

We learned that sustainable forestry is possible, but its success depends on a host of interconnected economic, social, and political factors; and that apparently small acts, like learning how to make handcrafted chairs with a mere handful of simple tools, can have ripple effects far beyond immediate communities.

At the start of the new millennium, and influenced in part by the then-ubiquitous presence of the word "sustainability," we proposed a television series that would respond to the question of how we could reinvent critical societal components (the food system, economics, cities, work) to make life more equitable and sustainable – for everyone. The audacious title of the series was *ReInventing the World*. In Brazil, Canada, and the U.S., we looked for thinkers, doers, and working examples of how systems could be designed differently.

In the *Food* program we went to Belo Horizonte, Brazil, where the municipal government asked a daring question: Why can't food be a human right? They then set about figuring out ways to make the city's food system more equitable. They supported local small farmers; lowered prices on fresh, local produce; and subsidized restaurants that offered fresh, local menus at a price even the very poor could afford.

In the *Cities* segment of the series, we began by asking a number of urban theorists to define sustainability. A planner in Portland, Oregon, gave one of the most concise responses: *"It means not eating your seed corn."* Writer Paul Hawken offered this thought: *"The problem with sustainability is that … it's not measurable … . Somehow sustainability implies that … we can balance our needs with the environment. But natural systems are enormously dynamic."*

Could it be that because of the fluid nature of life itself, sustainability is a process rather than a goal?

Our hope, and the reason we keep making films, is to create emotional connections to the issues we all face and the lives of the people who are most deeply affected by them. Changing attitudes, studies tell us, comes about less often from learning the facts than from hearing stories. By telling stories, by asking questions, by creating empathy, we hope to provoke reflection and to begin a process that encourages change. And because we favor questions, especially the most provocative, here's one to end on by writer Frances Moore Lappe: "Why are we as societies creating a world that we as individuals abhor?"

By our actions we will be judged: Is sustainability, in the end, really what we value most?

***

Heather MacAndrew *and* **David Springbett** *have worked together as life partners and creative co-workers for over 35 years, producing award-winning, internationally recognized documentary films and other media through their company Asterisk Productions. They are based in Victoria, British Columbia. Please visit them at www.asterisk.ca.*

Chapter 38
Conserving Energy for Tomorrow

Scott Tew

Just about everything I know about sustainability I learned growing up on my family's produce farm in Alabama. My grandparents, parents, and extended family taught me to love and respect the earth, the water, and the air. They taught me to work hard and pray for rain and sunshine – but not too much of either. And they taught me that natural resources are precious gifts that need to be nurtured, protected, and never squandered.

We took pride in our land and what it produced. For a time, if you lived in the Northeast and ate a watermelon before the Fourth of July, chances are it came from our family farm.

Mine was the first generation of our family to leave the farm, though by some measures I have not traveled all that far. I studied environmental science in college and spent the early part of my career performing environmental risk assessments. Later, I had the opportunity to help develop public policy on environmental issues and work for my company on projects connected with the National Oceanic and Atmospheric Administration (NOAA) and U.S. Fish and Wildlife Service, among other organizations.

> To me, sustainability is about the power of the word "and." As a society, we must find new and better ways to meet our needs today *and* conserve our limited natural resources *and* protect the planet for future generations.

Today, as head of the Ingersoll Rand Center for Energy Efficiency and Sustainability (CEES), my day job is helping organizations use energy wisely and reduce their carbon footprint. But I still occasionally get a little dirt under my fingernails tending my garden in North Carolina.

Part of our charter at the CEES is to make sure we have our own house in order by looking within Ingersoll Rand for ways to reduce energy consumption, improve environmental performance, and engage our employees.

For example, we are committed to reducing energy consumption and greenhouse gas emissions at Ingersoll Rand manufacturing locations by 25% from a 2009 baseline. As part of an overall energy-conservation program, we are using Trane eView™, a web-based software solution that enables employees to see how their actions affect energy consumption and make real-time decisions to reduce their energy use. Seven factories involved in a pilot program are on track to meet or exceed their 2-year interim goals, saving an estimated 40 billion BTUs and reducing greenhouse gas emissions by 31,000 metric tons.

Tapping into the energy and creativity of employees is a way that any organization can improve its sustainability. The Ingersoll Rand "One STEP Forward" program ties the company's sustainability objectives to employees' personal values and empowers them to take small but meaningful steps every day to reduce energy consumption and protect the environment.

The center is leading a company-wide effort to conduct product life-cycle assessments that examine – and ultimately reduce the size of – the environmental footprint our products create from the time they are designed and manufactured, during their long service life, and through to their ultimate disposal. We see opportunities to improve sustainability across our product portfolio.

We also are working to improve the way we package products. For example, packaging for one of our most popular personal security locks was redesigned this year using 90% less plastic.

The center recently received more than 300 submissions for a contest we sponsored to encourage designers, architects, and students to create a design for a simple, sustainable dwelling that can be built for $300. The challenge was originated by Dartmouth professor Vijay Govindarajan, who hopes to focus some of our best and brightest minds on finding ways to make affordable housing available throughout the world.

Engaging with customers and the community is a big part of my job at the center. I find that when I talk about "sustainable energy," many people envision hillsides covered with wind turbines, rooftops lined with photovoltaic cells, electric cars zooming along the highways, and airplanes flying coast to coast on biofuel made from algae.

Finding alternative sources of energy must continue to be a priority, but we also must face reality. The Institute for Energy Research says that fossil fuels meet 84% of current U.S. energy needs and will continue to be the centerpiece of the world's energy supply for generations to come.

> Sustainability presents enormous opportunities and challenges for this and succeeding generations. It is very exciting and gratifying to help my customers use energy more efficiently, reduce their environmental footprint, and accomplish their mission – all at the same time.

Emerging technologies get the lion's share of attention these days, but as someone who has spent much of his adult life helping people save energy, I am obligated to point out the critical role that energy conservation can – and must – play in creating a sustainable energy future for our nation and our planet.

With apologies to Benjamin Franklin, I would like to point out that "A kilowatt hour saved is better than a kilowatt hour produced." That is true no matter how efficiently that electricity is generated. When it comes to energy, the epitome of sustainability will be achieved when we successfully reduce the rate of growth in global energy consumption, generate only the energy we need, and produce the energy we use expending as few nonrenewable resources as possible.

But here is the catch. Fossil fuels are a finite natural resource, and the world's appetite for energy is growing at an annualized rate of about 1.4%, according to the U.S. Energy Information Administration (EIA). Using 2007 as a baseline, that means that world energy use will grow 49% by 2035 to about 739 quadrillion BTU. Not surprisingly, the EIA projects that energy demand in Brazil, Russia, India, China, and other developing countries will grow at a much faster pace, rising 118% during that same period.

Conservation is without question the world's most promising underutilized source of energy. In fact, research conducted by McKinsey and Company in 2009 concluded that the U.S. has the opportunity to reduce its annual nontransportation energy use by about 23% through improved energy efficiency. This would eliminate more than $1.2 trillion in wasted spending and reduce annual greenhouse gas emissions by 1.1 gigatons, which is the equivalent of taking every passenger vehicle and light truck off U.S. roadways.

Unfortunately, many people assume that "conservation" and "sacrifice" are synonyms. They may be willing to change their energy consumption patterns, as long as the changes they are asked to make do not affect their quality of life or limit their access to all the power-gobbling necessities and modern conveniences on which we have all come to rely. And I do not blame them.

But sacrifice does not have to be part of the energy conservation equation, as today's state-of-the-art, high-performance, green buildings clearly demonstrate. More and more new buildings are being designed using high-performance buildings principles. But the greatest potential benefit lies in using proven technologies and operating practices to improve the performance of existing commercial buildings, which account for close to 30% of the total U.S. energy consumption and a large share of annual greenhouse gas emissions.

The high-performance buildings concept takes a whole-building, whole-life-cycle approach that can reduce energy costs by 20–30% per year, according to the U.S. Green Building Council (USGBC). Substantial as they are, these energy savings just scratch the surface of the high-performance buildings approach's full potential.

High-performance buildings are designed and operated to meet specific standards for energy consumption, water use, building systems reliability, environmental impact, and other key performance measures. Operating standards are set, measured, and continually validated to deliver established outcomes within specific tolerances. What makes the high-performance buildings concept unique is that these standards link directly to an organization's primary mission and most important operational, financial, and customer-service objectives.

Hospitals set indoor air-quality standards that promote quality of care and support infection control objectives. Schools set acoustical standards that improve the learning environment by making it easier for students to hear what their teachers are saying. Offices and factories set reliability standards to prevent building system failures that could disrupt operations. Hotels set indoor environment specifications to ensure the health, safety, and comfort of guests and employees.

> Organizations and individuals should take a longer-range, full-life-cycle view when they make decisions about using natural resources and protecting the environment. Good stewardship is good business.

When the high-performance buildings approach is rigorously applied, a building becomes a value-producing asset that enables organizations and building occupants to be more efficient and more productive.

For example, a 2009 Michigan State University study found that workgroups moving into Leadership in Energy and Environmental Design (LEED)-certified offices achieved higher levels of productivity. The Center for Healthcare Design (CHCD) concluded that hospitals that do a good job monitoring and controlling their facility's physical environment achieve better patient outcomes. And numerous studies show that students in high performance schools tend to have higher test scores and fewer absences than those in conventional schools.

High-performance buildings can also help organizations improve their image and attract and retain customers, employees, investors, and tenants.

A CoStar Group study found that commercial buildings with LEED or Energy Star certification command higher rents, have fewer vacancies, and sell for premium prices.

Sound operating, maintenance, and services practices are essential to realizing the full potential of the high-performance buildings approach. The National Institute of Building Sciences (NIBS) estimates that energy and other operating costs represent 60–85% of a building's total life-cycle costs, eclipsing the amount typically spent on design and construction. ASHRAE has concluded that a poorly designed building operated and maintained effectively will usually outperform a well-designed building with poor operating and maintenance practices.

Intelligent service approaches that are holistic, technology-enabled, and knowledge-based ensure that a building's physical environment meets the mission requirements of the organization. Intelligent services technologies continuously monitor critical building systems and use sophisticated analytic and diagnostic tools to identify potential problems and enable building owners and operators to make informed, real-time decisions.

Given steadily rising energy costs, more stringent environmental regulations, and a global emphasis on competitiveness and productivity improvement, it is no surprise that improving building performance is gaining the attention of both the Oval Office and the corner office.

In 2011, the Obama Administration proposed a new Better Building Initiative to complement other government-sponsored energy and environmental programs. The administration's approach includes new tax incentives for improving building efficiency, more financing opportunities for commercial building retrofits, grants for universities and local governments to drive improvements, and training programs to create a pipeline of facilities professionals to support the emerging high-performance buildings environment.

Last year, my company sponsored a survey conducted by the *Economist* Intelligence Unit that found that 82% of senior leaders consider energy efficiency an important strategic priority for their organization, and more than three fourths said sustainability and efficiency will become even more important to them in the coming years.

Those findings jibe with the 2010 United Nations Global Compact–Accenture Chief Executive Officer Survey, which found that 96% of senior executives think sustainability should be fully integrated into their company strategy, and 93% see sustainability as essential to their organization's success.

This is clearly a period of transition in the evolution from yesterday's widely deployed technologies and practices to the evolutionary improvement offered by a high-performance buildings approach. Executives recognize the importance of improving energy efficiency and adopting sustainable business practices. But, at least for now, leaders appear hesitant to commit to major investments in energy conservation measures, even if those measures are projected to pay for themselves many times over a building's decades-long occupied life.

Without question, financial pressures are the primary factor in executives' reluctance to invest. In the wake of the most severe global recession in a century and a slower-than-expected recovery, organizations have tightened their purse strings and focused exclusively on opportunities with high return-on-investment potential and quick payback.

Further complicating matters, Accenture found that many CEOs believe investors have yet to fully recognize sustainability in their valuation models, which may discourage companies from making investments with longer-term payback. The *Economist* Intelligence Unit survey revealed that executives are generally skeptical about other organizations' claims about return on their energy-efficiency investments. They want more and better data before committing to large-scale sustainability initiatives.

Despite these challenges, it is clear decision makers in both the public and private sectors appreciate the importance of sustainability and recognize the value that leading-edge concepts such as high-performance buildings bring to the table. The expressed desire of some customer groups to do business with organizations that embrace sustainable business practices will likely provide additional momentum, as will greater clarity in the regulatory landscape.

Even though they are understandably reluctant to spend their scarce capital dollars in an uncertain economy, CEOs are beginning to recognize the potential of energy-saving investments to impact their bottom line. Ninety-one percent of respondents in the United Nations–Accenture CEO Survey said they expect to employ new technologies to help them reach their sustainability goals in the next 5 years.

Improvement in the economy, more and better real-world results proving the value of high-performance buildings concepts, and the ability of organizations to develop a clear and compelling business case will ultimately drive wider adoption of the high-performance buildings approach in both the public and private sectors.

The good news is that the technologies, tools, and world-class operating and maintenance practices exist today, making the value of high-performance buildings real and tangible. Early adopters are already reaping the benefits. The tipping point will come soon as more and more building owners and operators recognize that we do not have to wait for the "next big thing" to tap the enormous energy reserves that can be easily extracted from millions of underperforming buildings.

***

Scott Tew *is executive director of the Center for Energy Efficiency and Sustainability at Ingersoll Rand. He leads a global group of experts dedicated to integrating best practices for the long-term use of energy and other resources. He has a background in environmental science and ecology and extensive global public affairs experience. He serves on the national policy council for the Alliance to Save Energy and is an active member of the U.S. and India Green Building Councils.*

Chapter 39
The Sustainability of Ocean Resources

James Barry

Sustainability is all about time and perspective. I expect that most of us think of sustainability in relation to our own lifestyles over time scales of the years, decades, and perhaps a century or so that touch our lives, our children, our family. Over these time scales, natural resources we depend upon don't seem to change that fast or that much, and thus seem to be sustainable. If we take a somewhat longer view, however, sustainability refers to the use and stewardship of natural resources in a manner that allows society to benefit from their services today, without compromising their use for future generations. It is clear that humans are causing widespread changes in natural resources over land and the oceans. Many scientists consider the beginning of the Anthropocene (epoch of man) to be the sixth major mass extinction in Earth history due to the massive and continuing loss of biodiversity caused by our activities. The very resources we depend upon may not be functional or available for many generations into the future unless we change our policies and behavior concerning their stewardship. How did we get into this fix, and how can we change our course?

I'm not sure why, but most people seem to love the sea, regardless of where they live. Perhaps because the oceans are so vast, mysterious, and seemingly unchangeable – potentially terrifying, but often calming and inspiring, they have some allure for everyone. As a kid, I was lucky to surf, swim, fish, and explore the shores of California, and like most who spend time on the coast, developed a strong emotional connection to the ocean. So much so, it inspired my career in marine biology. I recall that I was generally concerned about ocean health and stewardship, but didn't really worry much beyond oil spills or toxic waste. We all knew the sardine fishery had crashed in Monterey and that overfishing and pollution were in the news, but seafood was usually available in the markets, and the waters were generally blue and clean. By and large, the oceans seemed relatively healthy and sustainable. I was far more worried about being attacked by a shark than how our fishing may affect them, and I expect that this view was fairly typical.

Fast forward several decades, the collapse of more fisheries, increased pollution, and add climate change and various other factors to the portfolio of human impacts on the oceans. The breadth and intensity of human-driven changes in many marine ecosystems appear to be approaching tipping points that may forever (at least for society) change these systems and the benefits we gain from them. Now, despite abundant evidence of the profound effects of human activities on ocean ecosystems, it seems that most of us are unaware of these changes and continue to view ocean life as healthy and enduring.

> Sustainability is all about time and perspective. Most of us think about sustainability in relation to our own life over the time scale of several decades that touch our lives, our children, our family. Over these time scales, natural resources we depend upon don't seem to change that fast or that much – and thus seem to be sustainable. A longer-term view is necessary.

The range of natural benefits we receive from the oceans is far broader than most of us realize. *Fish* – 20% of the protein consumed by society comes from the oceans. *Oxygen* – as we wander through each day, half the oxygen we breathe (every other breath) was produced by microscopic marine phytoplankton. *Clean water* – the oceans continually recycle our fresh water, then deliver it as rainfall over the land. *Climate control* – without the oceans, we'd be far too cold and far too hot. In addition to maintaining mild temperatures, the oceans also absorb about 30% of the carbon dioxide we emit as fossil fuel emissions, reducing greenhouse warming and further helping to stabilize our climate. *Coastline protection* – the coral reefs, mangroves, swamps, and similar coastal ecosystems provide natural coastline protection from storm waves and surges. The list of free stuff we gain from the oceans goes on to include the immense biodiversity of ocean life that is a storehouse of potentially important medicines for our future, recreational and spiritual experiences, and more.

Have you ever noticed that as technology always seems to make our lives better, much of what we use in nature, on land or in the oceans, is in worse shape than in the past? Every year, we seem to develop a plethora of cool new gadgets to improve our lives, but at the same time, most natural resources we use decline in some way. A frequent refrain about nature from old-timers is, "You should have seen it when I was a young." The fish were bigger and more plentiful, the reefs were pristine, tropical isles were unspoiled by plastics. Despite these trends, collapsing fisheries, coral bleaching, and other environmental damage, we forge ahead with the "good life" many of us lead,

> Every year, we seem to develop a plethora of cool new technologies to improve our lives, but at the same time, most natural resources we use are declining in some way. A frequent refrain about nature from old-timers is, "You should have seen it when I was a young."

and seem to worry little about the consequences for Earth's ecosystems. The sustainability of our "good life" somehow seems disconnected from the health of natural ecosystems. Why?

I think it has a lot to do with how we evolved as a species and as a society. Since the moment someone thought to sharpen a stick, technology has repeatedly saved us from famine by allowing us to exploit resources more efficiently or shift to new, more challenging alternatives. All organisms must exploit natural resources in one way or another to make a living. Through much of hominid history, we were similar to most animals, living as nomadic, hunter-gatherers of wild foods to support small populations. Technology has greatly reduced the manpower required to put food on the table. Toward the end of the Stone Age about 10,000 years ago, technology allowed the rise and expansion of agriculture, creating an ample and dependable food source. Consequently, population density increased around permanent villages, and our rates of survival and fertility increased. The epoch of humans kicked into gear.

As technology advanced, food production soared. Through the seventeenth to twentieth century, the British Agriculture Revolution improved technology for farming, including fertilizers, crop rotation, and other innovations that improved harvest, with reduced manpower. The release from intensive farming for part of the population led to advances in science, medicine, and others areas, resulting ultimately in the Industrial Revolution and modern society. The development of the oil industry in the mid-nineteenth century led to a drastic reduction in manpower requirements for food production. Synthetic fertilizer production in the early twentieth century reduced the dependence on natural fertilizers and stimulated food production, particularly during the Green Revolution of the mid-twentieth century, and greatly increased crop yields. To put the importance of the Haber-Bosch process of ammonia production in perspective, during the 100 years since, we have quadrupled our population and about one third of the world population now depends on crops produced using synthetic fertilizer.

All along the way, humans have become more and more efficient in exploiting natural ecosystems. While this has clearly increased our harvest of ocean resources, I think it has also become a sort of cultural Achilles heel. We expect that nature will continually rebound or that another fishery will be available to replace that which we've overfished. Throughout our history, this pattern has played out. When one stock was depleted, we had new technology

– bigger and more powerful boats fueled by low oil prices – that enabled the fishers to target unexploited species that were previously unfishable. Our approach for sustainability was to simply move on to the next available resource. In a sense, we have continued the nomadic theme of our ancestors, using technology to roam across the ocean seascape to hunt the next vulnerable stock. Crisis averted, at least for a while.

This technological fix has worked over and over in the past – will it work in the future? I doubt it. We've approached or reached key thresholds for ocean resource extraction. Wild fisheries harvest reached a plateau in the 1990s near 80 million tonnes per year, despite increasing effort. Recent increases in the harvest of ocean protein are due to the expansion of marine aquaculture, which will likely increase in the future, so long as we can develop sustainable methods. Fishing may currently be the largest impact of society on marine ecosystems, but is likely to be over-ridden by larger, more pervasive effects of our activities. We are causing fundamental changes in ocean conditions and ecosystems through our effects on climate. Carbon dioxide emissions are producing massive and rapid changes in climate that affect ecosystems on land and in the oceans. Coral reefs, for example, were lost during periods of rapid climate change in Earth history when carbon dioxide levels rose, and are threatened now by warming and ocean acidification driven by our carbon dioxide emissions. The reefs recovered following these ancient mass extinctions, but only after millions of years. Society can't wait that long. Unlike overfished stocks, which can be managed by thoughtful fishing practices, fossil fuel emissions and their ecological impacts require societal action on a global scale.

> The sustainability of our "good life" somehow seems disconnected from the health of natural ecosystems. Why?

The challenges concerning for effective long-term stewardship of natural resources are real and daunting. We hope to feed two to three billion more of us in the next century, while raising the standard of living for all. This means more protein, which will increase the demand for food production from both land and sea. Technology will undoubtedly play a key role, but we must move more rapidly toward sustainable practices for ocean resources. Recent progress provides hope for the future. New policies to address sustainability, including the new U.S. National Ocean Policy, mandating marine spatial planning, ecosystem-based management, and marine protected areas to restore and maintain the health of ocean ecosystems, will benefit society now and in the future.

We live in a remarkable and increasingly complex world, where a large portion of our population focuses not on food production, but on entertainment and leisure. We are addicted to oil, cars, smart phones, computers, electronic social networks, video, and a thousand other things that are replacing our experiences in nature. Our children often spend more time immersed in a digital world than a natural one. As we and our children spend less time in nature, on the shores, the waters, the forests, it is easy to lose our connections, understanding, value, and love for natural systems that truly sustain us in ways few appreciate. On the other hand, technology has also helped inform the public about nature in ways not possible in the past. Jacques Cousteau and his marvelous films on ocean life inspired millions who may never walk barefoot across a sandy beach or dive beneath the surface of the sea. A myriad of nature documentaries have followed to open natural world to the public, regardless of where they live. These are important tools that can promote an understanding of and value for natural systems, so long as we can achieve a balanced exposure for our children, using technology to help foster support for ocean health and sustainability.

Science must also assume a larger and more effective role in promoting stewardship by providing understandable and accurate information concerning the status of our natural resources. The public may appreciate and love the sea, perhaps inspired by nature films, but I don't think many people feel that they need the oceans for their very lives. The science community has a responsibility to inform the public on environmental issues. I think the marine science community deserves a fairly low grade in this and must become a more effective conduit for information about the status of ocean health and its importance for our lives.

When it comes down to making decisions about our future, our policies generally are driven by the will of the people. And we usually speak from our heart – we vote for what we value and love. A key challenge for the sustainability of ocean ecosystems and other natural resources will be to foster education, experience, understanding, and an appreciation of nature in our youth. Only then can we hope they will be better stewards than their ancestors.

***

James Barry *is a senior scientist at the Monterey Bay Aquarium Research Institute (MBARI), a nonprofit research institute in Moss Landing, California. As an oceanographer and marine ecologist, Barry has pursued a number of research interests at MBARI, supported mainly through the use of remotely operated vehicles diving in the deep waters off Central California. His research program focuses on the biology and ecology of marine animals. In particular, Barry studies the effects of warming, acidification, and declining oxygen levels in ocean waters - three consequences of fossil fuel emissions to the atmosphere - on the health of marine animals, and implications for ocean resources used by society. He has degrees in biology, zoology, and biological oceanography.*

Chapter 40

Will It Last? Will It Endure?

Andrea Coleman and Barry Coleman

In order for anything to endure, you have to understand the environment in which your work must endure – what are the external forces that could affect your work? And to ensure that you build the human, financial, and operational resources that will not fade or disappear over time. So, methods of internal maintenance must be built into your DNA from the start.

Furthermore, nothing will endure alone, nor is it worth anything surviving alone. The point of everyone's work in development is to make the whole work better. Your specific intervention or competence must contribute to making the whole work more effectively. So interdependence is the key to sustainability. Partnership in development means the maximizing and strengthening of the resources available.

These issues are plain for all to see on a daily basis in Riders for Health. Our *modus operandi* is focused on building local skills, building financial models that are durable, changing attitudes away from donor dependence – the greatest enemy of sustainability – and strengthening partners with our own core competence.

> Nothing will endure alone, nor is it worth anything surviving alone. The point of everyone's work in development is to make the whole work better. Your specific intervention or competence must contribute to making the whole work more effectively. So interdependence is the key to sustainability.

Barry Coleman, on a visit to Somalia in 1988, saw vehicles that were broken for lack of maintenance. No one had been trained to maintain them. So building skills was vital. Twenty years ago, the concept of "sustainable development" was still a long way from being the common phrase it is today. Our initial goal was not creating a solution for sustainable development, but about building an effective transportation infrastructure that would last. We focused on how we could sustain the running of a machine in rural Africa, because we knew the price to the community if we didn't.

Getting people to medical centers and delivering treatments to rural areas both rely on a regular transportation service. Children die of malaria because they could not be reached in time, or disease outbreaks spiral out of control because the community health worker had not been able to visit a particular village for months. Drugs and vaccines now exist to prevent many of the most widespread and deadly diseases, yet often they cannot reach the people who need them.

But there is a myth about transportation in Africa. Most people assume that in Africa's harsh terrain, vehicles will inevitably break down very quickly. Indeed, there is an overwhelming tendency of vehicles to break down in Africa after a very short proportion of their intended mechanical life. This is due to a widespread misunderstanding of the precise needs and nature of vehicles in hostile conditions, combined with a lack of vehicle maintenance.

Our innovation was the design of an appropriate, sustainable infrastructure in which to manage vehicles used in the harshest of conditions in Africa. Riders' emphasis is not simply on the provision of motorized vehicles, but on the management of these assets from the initial procurement all the way through user training, adequate servicing schedule by trained technicians, systematized spare parts procurement, provision of fuel, to resale at the optimum point.

In 2009, Gambia's Ministry of Health signed an agreement with Riders in which we owned and managed the entire fleet of outreach vehicles for the ministry. This public–private program, run on a not-for-profit basis, allows both parties to concentrate firmly on their core competency. It means that the health service knows that it will have reliable transport, allowing it to set its goals with conviction, and the savings made from better purchasing and from removing the costs associated with unexpected vehicle breakdowns can be better targeted at healthcare delivery.

> Our innovation was the design of a transportation infrastructure to help manage vehicles used in the harshest of conditions in Africa.

When working with Riders, partners see an increase to health worker productivity and an increase in coverage of key healthcare interventions. Health workers see a reduced time traveling and can spend more time in the community, seeing more people and visiting villages more regularly, improving both health impact and their own motivation. Finally, communities benefit from a much increased level of access to the vital preventive healthcare that will mean they can live disease-free.

If organizations can be assured that they will always have the transport to complete their work, and if they know how much it will cost in advance, it will mean that they can plan and budget with confidence. They will know that they will not have to return to donors or funders for more money to buy a new vehicle or to pay for emergency repairs. The focus is never on creating a sustainable organization for its own sake; it is to ensure sustainability for all.

> Anyone who is involved in an innovative solution will be judged by whether his or her ideas are sustainable. The sustainability we are working toward is the situation where the existence of reliable transport healthcare systems is taken for granted in rural Africa in the same way that it is taken for granted in Europe or the U.S.

What Riders for Health does is fundamental to the success of the partners we work with. In Lesotho, our work couriering diagnosis samples has dramatically reduced the time it takes to get patients onto antiretroviral treatments. In Kenya, we are helping community health organizations see four times as many families with regular visits. In Gambia, we have ensured that the Ministry of Health has all the vehicles it needs to reach its entire population. Across Africa, by putting in place reliable transport, we are helping to increase the public's trust in the health system. But without the work of our partners, our work would have no value.

Our goal was always to create a system for running reliable transportation for our partners. Creating mechanical sustainability for vehicles inevitably led us to create an organization that was robust enough to make this possible to achieve our goals. From an early stage, we knew that we would need diverse income streams to reduce our vulnerability on the changing focuses of global funders. The same cost-recovery model that helps to create the predictability for our partners and the sustainability of their service is the same structure that ensures that our organization is robust.

Creating a sustainable organization has never been the end in itself; it has only ever been the means. Anyone who is involved in an innovative solution will be judged by whether his or her ideas are sustainable. The sustainability we are working toward is the situation where the existence of reliable transport healthcare systems is taken for granted in rural Africa in the same way that it is taken for granted in Europe or the U.S.

***

Andrea Coleman *is chief executive officer of Riders and has guided the financial and advocacy development of Riders from its inception.*

Barry Coleman, *executive director of Riders, has nearly 20 years of experience in developing sustainable and sustained systems for managing motorized transport in hostile conditions. As co-founders of Riders, Andrea and Barry Coleman were selected to join the Schwab Foundation network of social entrepreneurs. They were also named as recipients of the Skoll Award for Social Entrepreneurship and as Ernst and Young U.K. Social Entrepreneurs of the Year.*

Chapter 41

The Holistic Enchilada: Moving Toward Food System Sustainability

Wayne Roberts

A sometimes beautiful, but often brutal, reality of food comes from its relationship to sustainability, which is simple and direct.

Individuals who don't eat anything will starve to death in about a month – not as quickly unsustainable as individuals who don't drink water or breathe air, more quickly unsustainable than individuals who don't have sex and reproduce, but quick and forceful enough to qualify as an individual relationship with a real and urgent need for sustainability. Their death will probably be defined as the outcome of a parasitic infection that worked havoc as a result of hunger, and will likely be described by friends as "tragically unnecessary," but will not likely be ascribed to unsustainability.

By comparison, individuals who mainly eat unsustainable amounts or types of food – food lacking necessary nutrients and food laden with harmful pesticides, germs, or manufactured goo, any of which undermines a person's ability to renew health – will likely die later in life from any number of incurable chronic diseases. Their death will be defined as the outcome of a specific disease, and described by friends and loved ones as "tragically young," but few would describe the cause of death as unsustainability.

> Holocaust survivor and psychotherapist Victor Frankl says the need for *meaning* is central to human survival, but meaning is something that humans create; it is not written into the script of the universe. So, my own meaning to sustainability? Actions of today that can make the life opportunities of our grandchildren, and the grandchildren of all species, healthier than those we grew up with.

By way of another comparison, individuals in societies that get their food from an unsustainable food system – a system that demands and degrades more resources than can be renewed or absorbed by natural systems of people, animals, plants, land, water, and air – can, if they so choose, stay alive until depleted natural systems around them crash. The cause of death at the time may be ascribed to an outbreak of rioting, or any number of contagious diseases that may or may not be acknowledged as related to food systems, or asphyxiation following a landslide of plastic bottles from a nearby recycling plant, or the inability of hospitals to cope with so many people suffering so many chronic diseases. Even at that late date, those who dare to name the cause of death as unsustainability will be dismissed as naysayers.

I have just retired after 10 years as manager of the Toronto Food Policy Council, a job I performed with as much relish, positive energy, and can-do attitude as I could muster. Some eco-minded types might chant, "Think globally, act locally," or "Eat locally, cook globally," but my personal mantra, which I largely kept to myself, was "Think darkly, act lightly." An insight I found very helpful was novelist F. Scott Fitzgerald's comment that "the test of a first-rate intelligence is the ability to hold two opposing ideas in the mind at the same time and still retain the ability to function."

Though the laws governing sad outcomes to unsustainable food choices are harsh and relentless, I think it is impossible to inspire people into sustainable food systems by scaring the daylight out of them. Even if it is possible to scare people into cutting back or stopping some bad things – drinking pop or eating potato chips, for example – very little that's good has ever come from fear-inspired revolutions.

I am hopeful and joyful as a person and food organizer because all the evidence – some of it cited in my book *The No-Nonsense Guide to World Food* – confirms two things: First, we already produce more than enough food to provide for everyone's need, and second, it costs us more to do the wrong things with food than it would cost to do the right things. We are shooting ourselves in the stomach because of our own inability to organize the abundance that surrounds us. The reason why sustainability in sustainable is that the

> The threesome of social, environmental, and economic sustainability has to become the "lens" through which all of our policies are viewed. Easy to say, but hard to overcome the resistance of bureaucratic empires that don't take kindly to any mandate beyond their particulars. This raises the need for popular education and engagement, which appeals to me because I think vibrant democracy is a touchstone of sustainability.

only barrier to it is our inability to organize ourselves out of a brown paper bag of our own making – aka narrow silos.

For all that most people think food is important, and might, albeit regretfully, agree with something like the three scenarios I lay out at the start, only the people of Norway, Wales, and Scotland have got their act together enough to have a food policy. Other countries may have an energy policy, an industrial policy (although that's a no-no among neo-liberals since the market should trump policy), a nutrition policy, a health policy, a food safety policy, a transportation policy, an antipoverty policy, a job strategy, a rural depopulation strategy, a trade policy, a waste management policy, a tourism policy, an air-quality policy, a community development policy, a compost policy, a water policy, a recycling policy, even a sustainability policy – I could go on and on until the cows don't come home – but very few have the moxie to say, "You can't have one of those policies unless you have a food policy because food choices will inevitably drive all those policies."

Once you have an overarching food policy that has some connection to any of the definitions of food security being floated around these days – my stab at one is something along the lines of "access for all to adequate, safe, nutritious, culturally appropriate foods essential to a healthy and active life, provided from a food system that respects social equity, environmental sustainability, organizational resilience, and community self-reliance" – you're off to the races.

The next thing we know, people working on garbage, hospitals, roads, community development, job creation – you name it – will start asking: What are we doing, and what can we do, to make progress on the sustainable food file?

If asking that question becomes a priority obligation of every government department, public and community agency, we're already halfway home. In food, to borrow a critical strategic insight from Jim Harris' book for organizations hoping to escape being blindsided, problem recognition must precede problem solving.

I actually believe problem solving becomes a piece of cake because most food problems are the result of narrow silos and the narrow thinking this incentivizes. If we simply stop financial and regulatory subsidies to foods that create health and sustainability problems, we can rely on the marketplace to put most of the bad actors out of business. Ending U.S. and E.U. financing of industrially produced grains, ending North American policies denying consumers information labels on genetically engineered corn and genetically engineered soybeans in most highly packaged junk foods, ending the Global North taboo against street vending of healthy foods that could allow small entrepreneurs to compete against junk-food chains – such low-cost measures would serve as appetizers for system-wide change that benefits the economy, health, and environment.

We do not have a food problem. On the contrary, we have food solutions waiting in the wings, held back by organizational problems. This is why the emerging movement for food system sustainability, based as it is on rock-hard foundations linking food simply and directly to sustainability, holds such promise.

> Stop government financial and regulatory subsidies to practices that are demonstrably unsustainable. Use the savings from such cuts to invest in sustainable practices that can become self-financing in the near future.

***

Wayne Roberts *is a Canadian food policy analyst and writer. He was manager of the Toronto Food Policy Council, a citizen body of food activists and experts that is widely recognized for its innovative approach to food security. He was also a leading member of the City of Toronto's Environmental Task Force. Since 1989, Roberts has written a weekly column for Toronto's* NOW Magazine, *generally on themes that link social justice, public health, and green economics. He has received the Canadian Environment Award for his contributions to sustainable living and the Canadian Eco-Hero Award presented by Planet in Focus. Roberts earned a Ph.D. in social and economic history from the University of Toronto and has written seven books, including* Get A Life!, Real Food for a Change, *and* The No-Nonsense Guide to World Food.

Chapter 42
Bringing Organizational Sustainability to Public Postsecondary Education

Christopher Hayter and Robert Hayter

Have you noticed that substantial cuts to state public postsecondary education budgets have become a familiar headline? If so, perhaps you've also noted subsequent announcements by public colleges and universities that they are cutting classes, services, and staff while casting a wistful eye toward a time when economies will improve and public funding will be restored.

We've seen a lot of arguments that attempt to stem this budgetary bloodletting through not-so-new messages: Public spending for colleges and universities should be maintained because of the historical role of these institutions in education, skill-building, and economic development.

There's no question that postsecondary education is important! But what has changed is the need for public colleges and universities to become sustainable; they will no longer be assured "regular" state funding – a trend columnist David Brooks calls "the new normal." What matters now is resource optimization: *how* public monies are spent and the outcomes that result.

> We think of sustainability as the strategic optimization of resources within complex systems.

States and public postsecondary systems will therefore need to work together to make sustainability a fundamental operating principle and consequently rethink the structure and performance of public colleges and universities. We focus specifically on state systems and policies because public institutions enroll a large proportion of students, account for a considerable share of state spending, and were established to meet specific state needs.

After World War II, public institutions adapted rapidly to the accelerating demand for postsecondary education, fueled by growing economies, skill needs, and generous federal aid programs. States built new and enlarged existing public universities – and created community college systems that continue to serve as both vocational training centers and gateways to four-year universities. By the late 1960s, the culture and management of public systems had crystallized around several key principles: keeping tuition low, maximizing student enrollment, faculty and student disciplinary specialization, and the familiar three-pronged mission of teaching, research, and community service.

Sustainability was not a relative priority among public postsecondary education systems – and the "business model" of most colleges and universities has changed little in past several decades. But demographic, technology, and budgetary trends highlight their unintended but inherent "waste." For example, while colleges and universities have received record numbers of applications and student enrollment is at the highest level ever, studies show stagnant or declining levels of degree completion, especially among specific demographic groups. Think of the student, faculty, and public resources that are lost when more than half of college freshmen don't make it to their sophomore year.

We've heard claims that this "waste" is a natural part of postsecondary education in the U.S., but many other countries have demonstrated that it doesn't have to be that way. For example, Korea has within a generation more than quadrupled its number of college graduates, earning the highest level of education attainment in the world among 25–34–year-olds while the U.S. has fallen to 10th. Furthermore, the Korean people spend far less on higher education as a percentage of GDP than their American counterparts.

Reasons for our diminished performance are myriad, including the poor preparation of public K–12 students, rapidly changing populations, and – yes – a harsh budget environment. But while some see these factors as crippling, we see them as further justification that sustainability must become a key operating principle for public postsecondary education.

To be clear, state and federal policy makers also contribute to the unsustainable nature of public postsecondary education. Not only does postsecondary education fall low on the relative priority list among policy makers, systemic change and resource optimization within postsecondary education is rarely part of a public agenda or lexicon.

Take the way states fund postsecondary education: Over time most states adopted funding formulas for colleges and universities determined by – among other factors – the number of students enrolled in classes early in the academic year, a de facto tuition subsidy. While this approach ensures some

> While interpretations of the word "sustainability" typically focus on natural systems, its application to organizations and institutions holds great promise. Institutions are important for establishing – and changing – norms and culture that govern our interactions with our surroundings. Over time, we must adapt to our environment – just as natural environments adapt to us – in a way that will hold promise for our mutual longevity.

funding equity among institutions, anyone who has taken a basic economics course knows that subsidies typically lead to waste and, consequently, substantial increases in cost over time – costs borne by taxpayers and students alike. A few states have experimented with new models but, by and large, public funding formulas have not been seen as a tool for driving desired outcomes in postsecondary education.

Our home state of North Carolina has funded postsecondary education through a formula and has done so generously; the state provides one of the highest per capita contributions in the country. North Carolina universities and community colleges are among the lowest *priced* in the nation but – ironically – some of the most expensive for taxpayers. According to 2008 U.S. Department of Education cost data, bachelor's and master's institutions in North Carolina *cost* an average of $14,440 per year, while the state's research universities cost an average of $20,010, the fourth and seventh most expensive in the United States, respectively.

And while society has yet to develop good metrics for postsecondary student learning, generous funding does not guarantee positive results. Well-respected public university systems in Texas, Colorado, and Florida, for example, have significantly higher graduation rates than those in New Jersey, Vermont, and North Carolina but at nearly half the cost.

Another example of the unintended profligate nature of postsecondary education is visible among public *and* private institutions where education "quality" is tied to a number of elements, including faculty "eminence," hypercompetitive admission standards, student amenities, and associated rankings in publications such as *U.S. News and World Report*. Prestige is of particular importance to colleges and universities and has long been determined by, among other factors, the number of articles that faculty publish in "highly ranked" academic journals and their capability to win research grants, typically funded by the federal government. This emphasis pervades most university promotion and tenure guidelines as well as decisions to hire new faculty.

The downside of becoming more prestigious, at least for public flagship universities, for example, is that they must – within a resource-constrained environment – focus more intently on their research mission at the expense of teaching and outreach. Again, there are many factors at play, but mounting evidence shows that these large public universities are consequently doing an increasingly poor job of graduating students and advancing social goals such as minority and low-income participation. These trends are often observed in large undergraduate classes: The average student often does not receive instruction or assistance directly from busy faculty members, who must increasingly worry about fundraising, conducting research, and publishing.

> The definition and application of "sustainability" needs to be expanded to include human behavior and organizations. Sustainability cannot be achieved without changes in culture and among organizations; we help define what environment means and consequently our relationship with it.

Sustainably meeting state goals within large, public flagship research universities is difficult at best given their sprawling size and complexity. However, their challenges are symptomatic of a larger "mission-creep" phenomenon trickling down into all public postsecondary institutions. Community colleges seek to add baccalaureate programs, while comprehensive universities are encouraged to become research universities. The desire for "upward mobility" is understandable. Faculty and administrators are aggressively pursuing their individual responsibilities to the best of their ability. But absent system- and institution-level focus on resource optimization, coordination, and common objectives, these individual actions collectively push institutions up the cost ladder: A year of classes at a research university is on average two to three times the cost of an equivalent year at a community college. More importantly, it distracts from an outcomes-based approach emphasizing student learning and skill-building among individuals of diverse ages and backgrounds.

Absent a clear "big picture" public purpose, supported by tangible outcomes and direction, funding cuts will continue just as reactionary responses make strategic restructuring even less likely. Even students and parents contribute to organizational inertia by rarely questioning the "efficiency" of postsecondary education so long as the *price* is low and they or their children can earn their degree, respectively.

It's clear that sustainability must become part of a renewed "social contract" between states and their public postsecondary systems. This starts with strong leadership; policymakers, education leaders, and the public must collectively ask and specifically answer the question: What are the people of the state getting in return for public funding of postsecondary education? A primary goal should be to ensure that taxpayers have access to increasingly advanced levels of low-cost, quality education with an emphasis on learning outcomes, not necessarily time in a classroom or how many articles faculty members publish. Other public goals may include aggressive policies to expand education attainment among minorities and economically disadvantaged and meet critical state workforce needs such as well-trained math and science teachers.

Addressing public goals will also require strategically rethinking the organization and operation of postsecondary education to ensure the right systems are in place to optimize resources – and to regularly reconfigure these systems to support that objective. At a state level, systems could begin by articulating a disciplinary and functional "division of labor" among colleges and universities, encouraging coordination, cooperation, and specific outcomes. A version of this idea was implemented decades ago with the California Master Plan; more recent and comprehensive versions include the North Dakota Higher Education Roundtable and the Virginia Restructuring Act.

State funding and policies will also need to be tailored to specific goals and institutions. For example, states may wish their public flagship universities to focus on research, advanced degree programs, and public service. To this end, states could deregulate these schools and contractually allocate funding, not by student, but by outcome such as the number of underserved students who successfully complete degrees, technical assistance provided to local businesses or communities, or the design of K–12 and postsecondary curricula.

Aside from system-level initiatives, public postsecondary education must also be rethought at the campus and classroom levels. Schools should rethink the delivery of instruction and closely tie it to learning outcomes. For example, instead of having senior research faculty "teach" large undergraduate vis-à-vis inexperienced graduate students, universities could create a career path – common among community colleges – for professional teachers who are responsible for student learning, with the former involved in labs and curriculum design. Furthermore, learning objectives should be tied to problem solving and real-world application. And institutions must find a way to recognize and reward faculty accomplishments among various types of scholarship and experiences while emphasizing teamwork and the fulfillment of system-wide learning objectives. Within this context, is tenure sustainable?

Even more dramatic is the "disruptive" potential of emerging online and e-learning systems that not only allow students the flexibility to complete coursework at any time in any location, but also provide ways to actually *measure* student learning. Talking about rethinking postsecondary education!

Sustainable public postsecondary education would be a complicated, never-ending venture requiring nonpartisan leadership, clear goals and metrics, and courage to question the status quo. But imagine the impact of transforming these institutions – and enabling technologies, policies, and funding – to improve learning outcomes and help meet other social and workforce needs of the state.

***

Christopher Hayter *is Director of Innovation and Sustainability at the New York Academy of Sciences, where he works with governments in the U.S. and abroad to evaluate and restructure postsecondary education systems.*

Robert Hayter *is an award-winning landscape architect and land planner as well as Vice Chair of the Board of Trustees for Sandhills Community College in Pinehurst, North Carolina. Both have experience as full-time college faculty.*

Chapter 43
Conservation Through Connections

Harvey Locke

I grew up in the Calgary-Banff area, where my family has deep roots in the Canadian Rockies. We spent much of our leisure time in Banff National Park, hiking and horse riding in the summer and skiing in the winter. We routinely saw wildlife, we drank directly from undammed streams, and I was reared on stories of family encounters with grizzly bears and wilderness. I thought this was normal. Then at 16 I went to college in the Swiss Alps, where all of that had been lost. Though the Alps are as beautiful as any place on earth, they had lost their ecological integrity and I felt the melancholy that is present wherever the natural world has been degraded by our species. I vowed that would never happen to my beloved Canadian Rockies. I felt strongly that at least in this one place, my home, the mountains should be protected.

> To me, sustainability means that we could get our act together to live comfortably and happily within the natural bounty of our incredible planet and share it fairly with the rest of life.

Fast forward 35 years: Today I work on the Yellowstone to Yukon Conservation Initiative, a grand vision to ensure connections for life among the great national parks and wilderness areas of the northern Rocky Mountains of Canada and the United States that seeks to protect at least half the region in a natural state. It has become interesting to people all over the world as a symbol of a new way to address conservation at scale. So I have had the honor of visiting every continent but Antarctica to talk about the idea and its application to local conditions. Through my travels I have learned to love lions and llamas, tigers and tuna, echidnas and elephants, and seashores and savannahs as much as grizzlies and glaciers. Now I think Yellowstone to Yukon should be the first example of the next relationship between humans and the natural world rather than being the best of the last. Instead of being a sort of anomalous living museum that clings to the last remaining fragments of Nature in affluent North America, it should be part of a global family of initiatives that will preserve the integrity of all life on Earth, in all its beautiful expressions, everywhere. So through the Nature Needs Half Initiative I am also working with people all over the world to have large-scale conservation become the norm for the whole planet, on land and at sea.

My journey over the last three decades has been one of constant learning and a search for ideas. About halfway along the way there emerged a concept so full of promise that it seemed to provide the answer: the idea of environmental sustainability. I remember how I learned about it. It was the late 1980s when I was a lawyer in private practice in a big law firm in Calgary and was about to be elected as the volunteer president of the Canadian Parks and Wilderness Society. I was grappling with the challenge of protecting wild nature while recognizing the need for humans to make use of the Earth's natural bounty. A friend told me that leading thinkers from many countries had gotten together through the United Nations' World Commission on Environment and Development to develop a global strategy to reconcile the two through a commitment to a new idea: *sustainable development*. The ideas were contained in a book entitled *Our Common Future*, which became known as the "Brundtland Report." I immediately ordered a copy, read and reread it, and lined up my thinking and actions behind it. Sustainable development to me was a rational and coherent strategy to reconcile humanity and Nature. I was so excited!

The publication of the World Scientist's *Warning to Humanity* in 1992 underlined the urgency for society to embrace the promise of sustainable development. It was signed by half of all living Nobel Prize winners, who warned:

> The underlying idea of sustainability is very appealing, but it is colored with disappointment because we are no closer to putting it into widespread practice than we were when the idea was first proposed.

> Human beings and the natural world are on a collision course. Human activities inflict harsh and often irreversible damage on the environment and on critical resources. If not checked, many of our current practices put at serious risk the future that we wish for human society and the plant and animal kingdoms, and may so alter the living world that it will be unable to sustain life in the manner that we know. Fundamental changes are urgent if we are to avoid the collision our present course will bring about.

They specified the hole in the ozone layer, species loss, climate change, deforestation, and overfishing, among their concerns. But I believed that we still had time to head off these problems through commitment to the practice of sustainable development.

Sustainable development also appealed to me because I am inherently an optimist. Despite humanity's many shortcomings, I like to believe that we can behave rationally and make the world a better place for our children and grandchildren.

Early signs in the sustainable development movement were really encouraging. The Montreal Protocol called on all countries to ban chloroflourocarbons (CFCs), which were widely understood to cause the hole in the ozone layer. This hole was allowing ultraviolent rays to burn through the atmosphere undampened, thereby blinding sheep in the Southern Hemisphere and raising the risk of skin cancer for all of humanity. Substitute chemicals were identified, CFC use was greatly diminished, and the ozone hole began to repair itself. My hopes were reaffirmed when in 1991 the leaders of the world gathered for the Earth Summit in Rio de Janeiro. They committed us to action to address the carbon emissions caused by humans that were changing the climate in dangerous ways (the Framework Convention on Climate Change) and committed their countries to ensuring the survival of all other species (the Convention on Biological Diversity).

In my native Canada, national and provincial roundtables were set up to bring all of society together in pursuit of sustainable development, and a National Green Plan was rolled out by the federal government. I helped a major oil company set up an environmental fund and worked with many of their senior executives to move ahead with what looked to me like a very genuine corporate and personal commitment to sustainable development. Some of my environmental colleagues were very skeptical. I barged ahead on the principle that we can't make the world a better place if we hold onto suspicions and fail to work with new allies. I was glad to give my time to it.

Then warning signs began to appear. I started hearing new twists on the idea through redefinition of the concept to "sustainable *economic* development." A person I knew well, who had strong industry, government, and roundtable connections, often said, "You cannot have a healthy environment without a healthy economy." One friend in the oil industry told me that, at his company, the senior executives saw the environmental concern as a fad that they just had to ride out. It turned out they were right. When the economy soured in the mid-1990s, the roundtables died, the oil company that I had given my time to abruptly changed its senior staff and cut expenses. The once society-wide enthusiasm for sustainable development supported by leadership from big business faded away. It had all been largely illusory.

It wasn't just big oil and other industries that undermined sustainable development. Attacks came from the left too. Postmodern ideas that made all coherent ideologies suspect because they were really tools of oppression used by elites to gain advantage over others gained support. A strange belief became widespread: that all action had to have local community support in order to be valid. It became odious to many on the left that a higher level of social organization might prevent locals, whether they be starving or just greedy, from rendering a species extinct or clearing the last old growth forest. It followed that any activity that served the development interests of a local community was automatically good and the environmental consequences were unimportant unless they were important to the locals. In the early twenty-first century, "Rio Plus 10," the global gathering on the environment in Johannesburg, South Africa, became focused on poverty alleviation, and this agenda also dominated the World Conservation Congress in 2004 so heavily that its theme was "Conservation for Development." Sustainable *human* development or sustainable *community* development was the focus. Nature's needs had become an externality.

> The condition of the Earth's natural resources is getting worse. We must stop defining sustainability as activity that supports economically or culturally driven bottom lines that are equal with nature for they are not: Human culture and economies are ultimately wholly dependent on the environment for their long-term success.

These perversions of the basic idea of sustainable development by both industry and the left have caused me great sadness. During their ascendancy, the climate continued to change while carbon emissions increased, the oceans continued to be emptied of fish, the rates of species extinction and at risk of extinction continued to worsen, wasted nitrogen runoff from farms created dead zones in the oceans, and deforestation ran rampant.

Most of the bad things the scientists warned us about 20 years ago are clearly upon us and worsening. Yet in 2011, a new preoccupation is gripping the world that threatens to distract us from a renewed focus on true sustainable development. It is called the "debt crisis." We have been so self-indulgent that we have spent and continue to spend more money than we take in. This has occurred across Western societies as a whole and at the level of the individual, the household, the community, the state, and the country. We have borrowed too much in order to feed our short-term desires and cannot pay our debts. Yet we still want all the services and benefits that borrowing gave us and simply refuse to either tax ourselves to pay for them or to reduce our demand for them. Until we stop spending more than the revenue we generate and return to living within our means and paying off our own debts, we will not solve the debt crisis. Bankruptcy and decline will follow with associated human misery.

The debt crisis is, of course, a mirror of our environmental problems and is a product of the same self-destructive behavior. We use up the Earth as though it were infinite, though we know better. We deplete and degrade soils and freshwater for short-term gain, we change the climate with abandon because we don't want to use less or more expensive energy, and we fish oceans to the point of collapsing the fisheries because we want to eat fish now. Unless we change our ways to live within the earth's means and restore the ecological debt we have created, we will face a horrific future of famine, human displacement, and resulting conflict.

Though my optimism has suffered along the road of experience, I have hope. I continue to believe in the promise of sustainable development. But for us to realize the promise of sustainability, humans must accept some basic and very simple realities. We are one species among many. We are a product of the Earth's biosphere and wholly dependent upon Nature (the interaction of the biosphere, the hydrosphere, the atmosphere, and the lithosphere) to furnish us with the air we breathe, the water we drink, and the food we eat. Without any one of these three things, we die. And we do not know how to make air, water, or food without Nature. Yes, we can manipulate Nature to increase yields, but we cannot make life, or air, or water. It is a time for us to accept these blunt truths, not to define them away in our own short-term self- interest.

Our economy and social development are wholly owned subsidiaries of the Earth's natural systems and wholly dependent upon them. It is not only ethical to keep all species and all the ecosystems they depend on healthy; it is entirely in our own self-interest. I believe, as do many scientists, that we should protect at least half of the Earth, both land and water, in an interconnected manner to enable all of life to continue to flourish on its own terms and so it can continue to support us. On the balance that is not protected, we should practice sustainable development.

I still have my original copy of the Brundtland Report. It describes sustainable development as meeting the needs of today without compromising the ability of future generations to meet their needs. This is still a robust and noble idea. We just need to get on with it by making our needs today live within the Earth's natural systems and to start repaying the ecological debt we have created. A hopeful future requires us to change and to embrace this way forward. It is as simple as that.

***

Harvey Locke *is strategic advisor for the Yellowstone to Yukon Conservation Initiative. Locke has been working to conserve Nature for over 30 years. He is well known for his work in the fields of large landscape conservation, national parks, and wilderness protection. He has been published, appeared in films, and spoken to audiences around the world.*

Chapter 44
Beyond the Status Quo: Catalyzing Sustainability in the Arts

Jane Milosch

Art enriches life and is an integral part of human existence. Art will always survive, but it will only thrive if we understand that it is more than an "afterthought" or "extra." It will flourish in the incubator of "sustainable development," in which art is situated within a broad context of exploration, exchange, and discovery. I would like to highlight two projects which I directed between 2008 and 2010, as well as some key programmatic and sponsorship ingredients, which reveal the dynamic role art plays in our lives through interdisciplinary study and shared experiences.

> Sustainability, from the perspective of an arts institution or an artist, means gaining a level of support that allows for the ongoing production as well as access to works of art. Bottom line: Funding and community building are absolutely critical.

The first project is a relatively new one at the Smithsonian, the Artist Research Fellowship, and the second an exhibition at the National Museum of Natural History in Washington, DC. Both projects draw on an interdisciplinary approach to the arts that the Smithsonian can uniquely offer, through collections and experts, along with a reliable level of financial strength. These are strong assets from which to generate new partnerships and collaboration at a variety of levels. Object-based study and experiential learning, together with meaningful exchange among the artists, scientists, and museum staff and visitors, increase our understanding and appreciation for art and underscore its connection with science and nature, society and life.

Since its inception in 2007, the Smithsonian Artist Research Fellowships (SARF) stipends have been awarded to 52 contemporary visual artists who take an interdisciplinary approach to subject matter, work in a variety of material and methods, and enjoy research and collegial exchange. SARF provides for a two- to three-month residency among the vast scientific, historical, and cultural resources of the Smithsonian Institution's diverse collections. Resident artists are also matched with experts on the Smithsonian staff. Artists are afforded with a highly focused period of time to closely investigate and study objects, events, or scientific phenomena related to their creative work. While other artist residency programs offer physical space and studios where artists can produce work, SARF provides an environment of discovery and contemplation. Fellowship recipients work in a variety of settings alongside experts in a wide range of disciplines – anthropology, astrophysics, botany, cartography, conservation, ethnography, ichthyology, and history.

Consider, for example, an exhibition, *The Bright Beneath: The Luminous Art of Shih-Chieh Huang*, which was recently on view at the National Museum of Natural History (NMNH). The work in this exhibition was inspired by the artist Shih-Chieh Huang's SARF residency, which involved the study of how organisms use bioluminescence and the evolutionary story behind its development.

> In the arts, sustainability is about human progress at a creative level. It encourages new ways of seeing and interpreting the world through a shared encounter. It also means going beyond the status quo and requires a reserve of resources to foster continued growth, innovation, and discovery.

During his residency, Huang worked closely with a research scientist and curator of fishes, Lynne Parenti. Their collegial exchange led Huang to create a series of kinetic sculptures for a site-specific art installation in Natural History's Ocean Hall. His interactive creatures – constructed from video cameras and monitors, motion and light sensors, fluorescent lights, electrical cords, fans, and plastic bags – almost seem to breathe and float as they expand and contract. As a result, the installation functions as a kind of animated interpretation of the scientific specimens on view nearby, wondrously enhancing our understanding of these deep sea creatures (Fig. 44.1).

While the SARF program has yielded great results and extraordinary collaborations, its future is still in a vulnerable position, especially in terms of dwindling budget and diminishing leadership. (Since 2009, the program has been administered by a committee.) A long-term plan for directing and funding the program needs to be developed, ideally with the goal to establish a named endowment in support of SARF. *Vision* and *funding*,

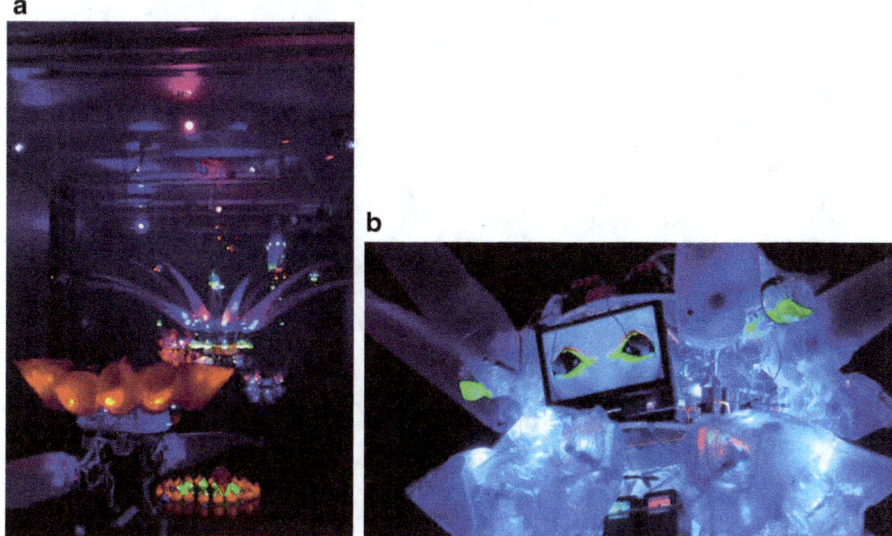

Fig. 44.1 (a and b) Installation and detailed views of Shih-Chieh Huang's sculptures, created from video cameras and monitors, motion and light sensors, fluorescent lights, electrical cords, fans, and plastic bags, from the exhibition, *The Bright Beneath: The Luminous Art of Shih-Chieh Huang*, at the National Museum of Natural History, Smithsonian Institution, Washington, DC, in 2011. This site-specific installation wondrously enhances our appreciation for art and science (Photos courtesy of the artist)

therefore, become a major factor for an institution committed to the sustainability of arts and promotion of collaborations to ensure the sustainability of arts and artistic minds.

In 2009, the exhibition department at National Museum of Natural History, inspired by the interaction and exchange between visiting artist fellows and their scientists, formed an Art and Science Committee to recommend art exhibitions and programs that offered new ways for their visitors to learn about science through art. At that time, I was invited to serve as an advisor to this committee, and later asked to co-curate Natural History's first contemporary art–science exhibition together with their marine science curator. The *Hyperbolic Crochet Coral Reef* project, a thought-provoking fusion of science, conservation, mathematics, and art, drew the participation of international and regional communities. Reef beauty, diversity, and environmental endangerment are highlighted by these vast, handmade, and ever-expanding crochet reefs. The project took inspiration from a discovery made by Cornell University professor Daina Taimina that scientific models of hyperbolic space could be created using crochet and illustrate the actual growth patterns of coral reefs. The traveling exhibition, *Hyperbolic*

Fig. 44.2 *The Smithsonian Community Reef* was on view at the National Museum of Natural History, from October 2010 through April 2011, as part of the traveling exhibition *Hyperbolic Crochet Coral Reef,* created and curated by Margaret and Christine Wertheim at the Institute For Figuring, Los Angeles (Photo by Eric Long, Smithsonian Institution)

Crochet Coral Reef (HCCR), created and curated by Margaret and Christine Wertheim at the Institute For Figuring (IFF), Los Angeles, invites volunteers at each of the exhibition venues to crochet and exhibit their own community reef alongside the traveling reefs, and led to the creation of *The Smithsonian Community Reef.* The Art and Science Committee agreed that this exhibition and community project would be a great way to initiate a contemporary art–science program, meeting the criteria for interdisciplinarity and offering the potential to attract a new and diverse audience to NMNH.

The project's community outreach component played a major role in its success. Through workshops hosted at the museum, online exchange, regional yarn shops, and community centers, 800 individuals hand-crocheted and donated more than 4,000 individual "pieces of coral" to the *Smithsonian Community Reef* (see Fig. 44.2). The *community-building aspect* of the *Smithsonian Community Reef* heightened an appreciation for the handmade and served as a reminder that individual actions can have a powerful, collective impact on our environment.

While the initial funding for the exhibition was covered under the museum's exhibition budget for traveling shows, it was necessary to raise the funds to make the community participation possible. Three dynamic sponsors – Quiksilver Foundation, The Embassy of Australia, and the Coral Reef Alliance (CORAL) – who had shared interests in the goals and success of the exhibition became enthusiastic partners. All three lent not only financial support but also educational and program resources to help promote the project. Also critical was hiring a project coordinator with expertise in fiber arts and with some knowledge of science and environmental issues. The coordinator became a lynchpin to the project's success, building and managing the relationship between the museum and community and ensuring the aesthetic excellence of the community reef.

Achieving a sustainable audience level has long been a goal of museums. NMNH successfully created a dedicated following of visitors and participants who enjoyed the collaboration with the museum and project. This volunteer group, ready and willing to be mobilized for a similar type of project, eagerly awaits, but there is no mechanism or financial support to engender another collaboration within the current educational structure. NMNH would like to continue this outreach model, but new funding sources and community-development methods need to be ensured for this to become sustainable.

> We need to set our egos aside and collaborate across disciplines as a first step toward sustainable development. The more connections we see and make, the more alive and aware we become.

Critical components for sustainable development of the arts are funding and visionary leadership. When the leadership of large organizations becomes too entrenched, it often short-circuits innovation and discovery. The work of a museum is about short- and long-term cultural investment and requires leadership that remains supple and open. While the arts and the government are not ideal partners for sustainable development – government tends to operate "inside the box," while art thrives outside it – public funding offers a stable base with which to attract private funds. But in order for this to happen, it takes visionary leadership that values what is already in place while looking to the future. The importance of understanding the past and the ability to reframe, again and again, leads to new partners and sponsors. Too often new leadership imposes an entirely new game plan or a top-down initiative without soliciting internal expertise and looking toward long-term outcomes. Museums and other cultural institutions need beware of bureaucratic entrenchment and an attachment to legacy models of success. Sustainable development requires current leaders to identify and cultivate the next generation of leaders, one with diverse skills, through mentoring and offering on-the-job leadership experience.

In recent years, the most successful attempts to attract new audiences and financial support have involved combining contemporary art with public programs, implementing interdisciplinary approaches to exhibitions, and developing community-based projects that generate meaningful connections to the institution. In order to reach a level of sustainable development, we also need to consider the importance of collaborating with creative and dedicated patrons of the arts. And this is why I am advocating for something like a "Smithsonian Dream Team for the Arts," a group of dynamic and committed individuals who envision the arts going beyond sustainability. When art is integrated into our understanding of culture and society, it helps us see possibilities. Artists and creative people in a variety of disciplines and fields help us to become "unstuck," to consider new ways of thinking and approaching problems and opportunities – they are catalysts for change and sustainability.

***

Jane Milosch *is director of the Provenance Research Initiative in the Office of the Under Secretary for History, Art, and Culture at the Smithsonian Institution, where she has also served as senior program officer for the arts. Previously she was a curator at the Renwick Gallery, Smithsonian American Art Museum, Washington, DC, Cedar Rapids Museum of Art in Iowa, and the Detroit Institute of Arts. She was a Fulbright Scholar in Germany. Her research interests include American craft, decorative arts, and contemporary art.*

Chapter 45
Historic Preservation: The *Real* Sustainable Development

Donovan Rypkema

In a great Tom Robbins book, *Skinny Legs and All*, one of the characters is an extremely erudite can of pork and beans. At one point, Can 'O Beans remarks,

> Imprecise speech is one of the major causes of mental illness in human beings. The inability to correctly perceive reality is often responsible for humans' insane behavior. And every time they substitute a ... sloppy slang word for the words that would accurately describe a ... situation, it lowers their reality orientations, pushes them farther from shore, out onto the foggy waters of alienation and confusion.

The phrase today that is the best example of imprecise speech is *sustainable development*. If we listen to environmental activists, sustainable development is saving the rainforest and the habitat of the snail darter. If we listen to the U.S. Green Building Council, sustainable development is solar panels and waterless toilets. Saying "green buildings" is a synonym for sustainable development is equivalent to saying going to the dentist is a synonym for healthcare – an important component, yes, but far from the whole story.

> Of the dozens of definitions I've read for sustainable development, I've yet to find one better than what emerged from the Brundtland Commission nearly 30 years ago: "the ability to meet our own needs without prejudicing the ability of future generations to meet their own needs."

We don't yet get it in the U.S., but the rest of the world is beginning to. The international framework for sustainable development certainly includes environmental responsibility but also economic responsibility and social responsibility.

That creates three important nexus: For a community to be viable, there needs to be a link between environmental responsibility and economic responsibility; for a community to be livable, there needs to be a link between environmental responsibility and social responsibility; and for a community to be equitable, there needs to be a link between economic responsibility and social responsibility.

It is not the Sierra Club or the EPA or the Nature Conservancy who are leading the way in comprehensive sustainable development. The real sustainable development is done through historic preservation.

How does historic preservation fit the definition of sustainable development?

On the environmental side, the green building approach focuses almost entirely on the annual energy use of a building. But the energy expended to build the structure is 15–30 times the annual energy use. This is called *embodied energy* and is defined as the total expenditure of energy involved in the creation of the building and its constituent materials. None of the measurements of annual operating costs account for this embodied energy.

Certainly there can be improvements in the energy efficiency of some historic buildings – and preservation architects and conservationists are developing methods to make those improvements without sacrificing the character-defining features of the building. But because of the embodied energy in the structure, a 100-year-old building could use 25% more energy each year and still have less lifetime energy consumption than a building that only lasts 40 years. And a whole lot of buildings being built today won't last 40 years.

And then there is the environmental impact of demolition. Landfill is increasingly expensive in the U.S. in both dollars and environmental quality, and a third of everything dumped at the landfill is construction debris, including the remnants of razed historic structures. Americans diligently recycle their Coke cans. But demolishing one small masonry commercial structure – 2 stories, 25 ft wide and 120 ft deep – wipes out the entire environmental benefit of the last 1,344,000 aluminum cans that were recycled. At most, perhaps 10% of what the environmental movement does advances the cause of historic preservation. But 100% of what the preservation movement does advances the cause of the environment. You cannot have sustainable development without a major role for historic preservation, period.

How does historic preservation advance economic responsibility? In a multitude of ways.

- Jobs. In Georgia, for example, $1,000,000 spent rehabilitating a historic building creates nearly 15 more jobs and generates over half a million more in household income than manufacturing a million dollars of automobiles. (*Good News in Tough Times: Historic Preservation and the Georgia Economy*, 2010).
- Household income. In Connecticut, every $1,000,000 invested in historic building rehabilitation ultimately means $800,000 in the pockets of Connecticut workers. (*Investment in Connecticut: The Economic Benefits of Historic Preservation*, 2011).
- Downtown revitalization. Virtually every sustained success story in downtown revitalization in America has had historic preservation as a key component of the strategy. The very expensive failures in revitalization efforts have all had the demolition of historic buildings as a priority.
- Property values. In study after study, properties protected by (and regulated by) local historic districts have appreciated at rates faster than similar non-designated neighborhoods, and faster than the local market as a whole.
- Property value stability. In analyses in Pennsylvania and Connecticut, foreclosures in historic districts during the current real estate recession occurred at half the rate of comparable neighborhoods.
- Small business incubation. It is not an accident that the creative, innovative, small startup business isn't at the regional shopping center or the corporate office park. They can't afford the rents there. Older and historic buildings provide a natural incubation for these businesses, usually with no assistance or subsidy of any kind.
- Infrastructure savings. Preservation projects save 50–80% in infrastructure costs compared to new suburban development.
- Economic stimulus. Over the last 30 years, the historic tax credit program of the federal government has created 1,815,000 jobs at a cost of $9,222 per job. While not surprisingly the largest share (28%) are in the construction trades, manufacturing (20%), services (18%), and retail (15%), each saw major job creation. The White House reports that the Stimulus Plan, on the other hand, has created jobs as a cost of $445,183 per job.

> Sustainable development is about responsibilities – environmental responsibility, economic responsibility, and social responsibility. In the twentieth century, most major movements were about rights – such as civil rights, women's rights, rights of free speech, political rights, animal rights, prisoners' rights. But the exercise of rights without the corresponding responsibilities is what has gotten us into the economic, environmental, and social chaos we find ourselves in.

- Tourism. Heritage visitors stay longer, visit more places, and spend more per day than other tourists. Therefore, the per-trip expenditure has as much as 2½ times the local economic impact.

One common response is, "But the jobs created aren't sustainable, right? Once the building is renovated, the jobs are gone." This is equivalent to saying, "Making solar panels doesn't create jobs. Once the solar panel is manufactured, the jobs are gone." But the assumption is that when one solar panel is made, the worker will move on to make the next one. Likewise, once a building is renovated, the worker will move on to the next one. In fact, because most building components have a life of between 25 and 40 years, a community could commit to rehabilitating 2–3% of its buildings per year and have perpetual employment in the construction trades – the ultimate in sustainable economic development.

Finally, historic preservation *is* social responsibility. Historic buildings are the physical manifestation of memory. Historic preservation's role in helping us understand who we are, where we have been, and where we are going is central to the social component of sustainable development. The sociologist Robert Bellah wrote, "Communities ... have a history – in an important sense they are constituted by their past – and for this reason we can speak of a real community as a 'community of memory', one that does not forget its past."

> The environmental movement in the U.S. has co-opted the phrase "sustainable development" as if it were exclusively about the environment. It is not. Until the phrase is understood to be composed of these three co-equal components, we will have a truly unsustainable "sustainability" movement.

If environmental activists don't yet get the connection, the Inter-American Development Bank does, noting, "As the international experience has demonstrated, the protection of cultural heritage is important, especially in the context of the globalization phenomena, as an instrument to promote sustainable development strongly based on local traditions and community resources."

The closest to a comprehensive sustainable development movement in the U.S. is known as "Smart Growth." The Smart Growth movement has established an excellent set of principles:

- Create range of housing opportunities and choices.
- Create walkable neighborhoods.
- Encourage community and stakeholder collaboration.
- Foster distinctive, attractive places with a Sense of Place.
- Make development decisions predictable, fair, and cost-effective.

- Mix land uses.
- Preserve open space, farmland, natural beauty, and critical environmental areas.
- Provide variety of transportation choices.
- Strengthen and direct development toward existing communities.
- Take advantage of compact built design.

Great list. But if a community did nothing but save its historic buildings and neighborhoods, every Smart Growth principle would be advanced.

Every fifth grader in America is taught that saving the environment means *reduce, reuse, recycle*. Rehabilitation of historic buildings reduces the demand for land and new materials; reuses energy embodied in the existing materials, the labor, skills, and the urban design principles of past generations; and recycles the whole building. In fact, historic preservation is the ultimate in recycling.

I'm not against LEED certification. I'm also not against solar panels or storm windows. But what does a solar panel or a storm window recycle? Nothing. What does a solar panel or a storm window reuse? Nothing. They both might reduce energy use, but much of that savings in offset in the energy used to build the damn thing: in the case of an aluminum storm window, 126 times more energy than in repairing an existing wood window.

We can preserve wetlands and be environmentally responsible ... but have no effect on economic or social responsibility. We can teach local history in the elementary school and be socially responsible ... but have no effect on economic or environmental responsibility. We can have an equitable tax system and be economically responsible ... but have no effect on cultural or environmental responsibility.

It is only through historic preservation are we doing all three simultaneously.

The demolition of historic buildings is the polar opposite of sustainable development; once they are razed, they cannot possibly be available to meet the needs of future generations. In a world filled with rights movements, historic preservation is a responsibility movement – responsibility toward our environment, responsibility toward our economy, and responsibility toward our social and cultural lives.

So go and buy a solar panel and a waterless toilet if it makes you feel good. But if you really want to be part of sustainable development, rehabilitate a historic home.

***

Donovan Rypkema *is principal of PlaceEconomics, a real estate and economic development consulting firm in Washington, DC. Rypkema is recognized as an industry leader in the economics of preserving historic structures and has performed real estate and economic development consulting services throughout the U.S. for state and local governments and nonprofit organizations, with interests in a broad range of properties, from National Historic Landmark Structures to Main Street commercial centers. He received an M.S. in historic preservation from Columbia University. He is author* of The Economics of Historic Preservation: A Community Leader's Guide.

Chapter 46

Bending Toward Justice: The Search for Sustainable Energy

Michael Brune

During the weeks leading up to a decision by the U.S. government over whether to allow a Canadian company to construct a 1,700-mile pipeline to carry tar-sands oil from Alberta to the Gulf of Mexico, environmentalists and clean-energy advocates mounted a huge opposition campaign. But they were joined in their resistance to the Keystone XL pipeline by two public figures that you might not immediately associate with energy policy: the Dalai Lama of Tibet and Archbishop Desmond Tutu of South Africa.

What motivated those two revered figures to move to speak out against an oil pipeline? The answer to that question lies at the heart of what "sustainability" means when we evaluate energy solutions.

> Sustainable systems are in balance. They maintain equilibrium. If they don't, they tend to crash. Unfortunately, perfect equilibrium is incredibly difficult to engineer.

There is a tendency to think that "sustainability" simply refers to the long-term viability of a solution. In that sense, the burning of fossil fuels for energy is obviously not sustainable because the natural resources themselves – coal, oil, natural gas – are finite. Our planet has only so much of them, and when they run out, or become too difficult to extract, that's it. Intellectually, each of us knows this will happen, but the assumption that it will happen only sometime in the distant future diminishes our sense of urgency.

Frankly, that's not working for us. But there's a different way to look at sustainability that can better inform our energy choices: Sustainable systems are in balance. They maintain equilibrium. If they don't, they tend to crash.

Unfortunately, perfect equilibrium is incredibly difficult to engineer. But aiming toward balanced energy solutions can help us make better choices. And I'd say that the opposite of a balanced solution is usually one with inequities that lead to injustice. So let's ask ourselves: Are we meeting our energy needs in ways that consider the present *and* the future, the wealthy *and* the poor, the needs of the individual *and* the welfare of the community?

I can be as inspired as anyone by the balance exemplified by a healthy ecosystem, but I studied finance and economics in college. And if I were auditing the books for energy policy, I'd have to say they don't add up. Too often, we've embezzled from the future, robbed the poor, and acted irresponsibly.

At the international level, the failure of the U.S. to take its share of the responsibility for the global problem of carbon pollution is a clear example of short-sighted self-interest blinding us to our moral obligation, not to mention the economic opportunities inherent in clean energy, which employs more people than fossil fuels, reduces healthcare costs, and creates genuine energy security. Our country, which is generating by far the greatest amount of carbon pollution per capita, has failed to act – sowing seeds of resentment among the international community. That inequity fosters a sense of injustice that cannot lead to a balanced, sustainable solution to the world's problems.

> As a parent, I am deeply concerned about the future we are creating for our children. Additionally, a fundamental sense of fairness and justice demands that we act responsibly to preserve and protect this planet for all humanity – not just a privileged few.

At the national level, the scales of U.S. energy policy have been tipped by an entrenched and financially powerful fossil fuel industry that is fighting to prolong its outsized profits and influence. One of the things that excited me about joining the Sierra Club as executive director in 2010 was its extraordinarily successful campaign to redress that imbalance in our national energy policy by taking coal-fired power plants out of the equation. Why is it so important to get America (and the world) to stop burning coal? It's not because coal is a nonrenewable resource, but because the harm it does far outweighs the benefit of the energy it provides. That harm is measurable – in terms of public health, climate disruption, environmental justice, and the chilling effect on development of other, less-harmful energy technologies.

> The politicization of energy issues has distracted from the universal human values that should be the basis of an equitable and sustainable energy policy.

But since coming to the Sierra Club, I've also had reinforced for me the lesson that we must be rigorously honest in our evaluation of all energy sources and how "sustainable" they really are. Coal is dirty and causes great harm. Natural gas, on the other hand, burns "cleaner" than coal or oil – but still contributes significantly to greenhouse gas emissions and also threatens our air and water. We must prioritize the rapid expansion of solar, wind, and geothermal energy, while extracting the most gain from stronger investments in energy efficiency. We may then include some natural gas in the mix to replace coal-fired power plants in the short term, but we won't be getting closer to the kind of balanced, sustainable solution we need if we don't minimize that gas use and if we fail to drill the gas as responsibly as possible (something that, so far, we certainly can't take for granted).

Let's look closer at wind and solar. We won't run out of either resource – at least not for a few billion years. It's true that wind and solar have the potential to provide a long-term, sustainable energy supply, but even these resources can have issues. Large-scale solar developments, for instance, have to be built somewhere. One person's "desolate" desert landscapes might be another's unique wilderness resources, home to endangered or threatened species. Wind turbines should be designed and sited responsibly to minimize wildlife fatalities. The downsides of both – and all forms of large-scale energy facilities – can be reduced through a shift to smaller-sized facilities – on the rooftops of homes and warehouses and in abandoned lots.

The Sierra Club's founder John Muir famously observed, "When we try to pick out anything by itself, we find it hitched to everything else in the universe." That's exactly what makes our universe such a complicated, frustrating, and wondrous place. Humans, clever though we are, will never have all the answers for how to make it work the way we think it should. There's no owner's manual. No help menu.

But the universe has another quality, beautifully expressed by Dr. Martin Luther King, Jr., that can help point us in the right direction. "The moral arc of the universe is long," he said, "but it bends toward justice." That is why I am sure that, in the long run, the U.S. will come to its senses on climate disruption, that the world will turn its back on coal, and that truly sustainable energy policies will ultimately prevail.

***

Michael Brune *is executive director of the Sierra Club, the largest and most influential grassroots environmental organization in the U.S. He is particularly interested in promoting programs that link the Sierra Club's traditional protection of wild places, including national parks, to urgently needed climate change solutions. He previously served as executive director of the Rainforest Action Network and holds degrees in economics and finance from West Chester University in Pennsylvania. Brune is author of* Coming Clean: Breaking America's Addiction to Oil and Coal.

Chapter 47
Afterword

M.S. Swaminathan

I am very excited about *Practicing Sustainability*. This book's core strength stems from its innovative approach – it includes many perspectives that are utterly new and sometimes controversial. *Practicing Sustainability* is not only the need of the hour but also a luminous work bolstered by its restless, relentless, provocative, and groundbreaking variety and freshness of views – attributes that verily underpinned Charles Darwin's own discoveries. I very much hope this work will serve to inspire a new generation of innovative practitioners in sustainable development.

"Variety is the spice of life" is a common saying. Variation, in point of fact, is necessary for natural selection to occur. We can augment natural selection processes – which arise naturally from existing variations – to help provide for a more sustainable environment. For example, our research team at the M.S. Swaminathan Research Foundation in Chennai, India, has been able to develop salt-tolerant varieties of rice by transferring genes from the mangrove species. Such halophytic – salt-tolerant – plants can help to launch a seawater farming movement along our vast coastlines. Agricultural and forestry systems involving the cultivation of mangroves and other salt-tolerant shrubs and trees, together with the cultivation of farmed fish, will open up new livelihood opportunities to coastal communities.

Artisanal or small-scale fisheries can become economically and technologically attractive by introducing cell phones carrying data on wave heights and the location of fish shoals. Also, the seawater farming methods can help to increase yield and income from coastal aquaculture. By linking the Darwinian concept of variation with the powerful tools of molecular genetics, we can not only safeguard our food security, but also strengthen the ecological security of coastal areas, and the livelihood security of coastal communities.

Experts and administrators of every stripe participate in thousands of conferences annually related to sustainability issues. Lofty declarations are made, but these words are not followed up with action. This book provides a realistic attempt to assess why there has been little real progress in this area, and to help provide a solid foundation to move forward.

In sum, "sustainable development" is a term that is frequently misunderstood, misinterpreted, misapplied, and misused. But the sustainability movement itself is vital and confronts extraordinary challenges. If we are to provide an intelligent birthright for generations to come, now is the time to create new pathways for balancing human needs with sound environmental practices. It is clear there will be more disillusionment to come unless there is serious effort to understand our goals for sustainability, and to approach those goals with well-reasoned practicality. That is what *Practicing Sustainability* is all about.

GPSR Compliance

The European Union's (EU) General Product Safety Regulation (GPSR) is a set of rules that requires consumer products to be safe and our obligations to ensure this.

If you have any concerns about our products, you can contact us on

ProductSafety@springernature.com

In case Publisher is established outside the EU, the EU authorized representative is:

Springer Nature Customer Service Center GmbH
Europaplatz 3
69115 Heidelberg, Germany

www.ingramcontent.com/pod-product-compliance
Lightning Source LLC
LaVergne TN
LVHW010338260326
834688LV00036B/769